Autodesk AutoCAD 2013 Practical 3D Drafting and Design

Take your AuotoCAD design skills to the next dimension by creating powerful 3D models

João Santos

[PACKT]
PUBLISHING

BIRMINGHAM - MUMBAI

Autodesk AutoCAD 2013 Practical 3D Drafting and Design

First published: April 2013

Production Reference: 1180413

Published by Packt Publishing Ltd.
Livery Place
35 Livery Street
Birmingham B3 2PB, UK.

ISBN 978-1-84969-935-8

www.packtpub.com

Cover Image by João Santos (jsantos@qualicad.com)

Credits

Author
João Santos

Reviewers
Décio Ferreira

Filipe Vila Francisco

Acquisition Editor
Andrew Duckworth

Lead Technical Editor
Susmita Panda

Technical Editors
Sayali Mirajkar

Kaustubh S. Mayekar

Soumya Kanti

Copy Editors
Aditya Nair

Laxmi Subramanian

Alfida Paiva

Project Coordinator
Joel Goveya

Proofreader
Claire Cresswell-Lane

Indexer
Rekha Nair

Production Coordinator
Nilesh R. Mohite

Cover Work
Nilesh R. Mohite

About the Author

João Santos is the manager and main instructor at QualiCAD (www.qualicad.com), one of the most important Portuguese ATCs (Authorized Training Centers), based in Lisbon. With a degree in Mechanical Engineering, he has been teaching AutoCAD for more than 25 years now and is an ATC coordinator for almost 20 years. He is an AutoCAD 2013 and 3ds Max 2013 Certified Professional User and Instructor. He is also the Portuguese instructor in these technologies with more students. He is the author and co-author of more than 40 AutoCAD and 3ds Max books, all written in Portuguese.

First I would like to thank my family for all the support and guidance. Not less important are all my students, friends, readers, and colleagues for continuous questions, feedback, suggestions, and basically, shaping my career. Special thanks to friends and experts Décio Ferreira and Pedro Aroso for development ideas and 3D models. And I would also like to express my gratitude to the entire team at Packt Publishing for this opportunity and collaboration.

About the Reviewers

Décio Ferreira is an architect, actually working on PFarquitectos, and began working with Autodesk software early. He started his career as an architect using AutoCAD ® R12 and in 2000 began working with three-dimensional tools, such as ADT 2.0 and 3.0 3DS Viz. He adopted Autodesk ® Revit ® Architecture as his main working tool quite early. He is a Certified Instructor (41,920 Approved Instructor) by ATC and ACC, EdiCad Computação Gráfica e Imagem Lda, Portugal. Also, he is part of the technical team CPCis (VAR) as Post Sales Engineer. He is a certified professional in Revit Architecture 2011, 2012, and 2013 and AutoCAD 2011, 2012, and 2013, and has several hours of experience in training, consulting, and implementation of Revit technologies in several Portuguese offices. He is also the moderator of the forum Revitpt (http://www.revitpt.com/) and Revit developer blog in Portugal (http://revit-pt.blogspot.pt/).

Filipe Vila Francisco is a Senior CAD Designer and Certified Instructor. He has been working with AutoCAD since Version 12 and has over 15 years of using Autodesk software.

He is also an expert in AutoCAD and an AutoCAD Certified Associate and Professional since 2008. He is the author of the blog, CAD4MAC and has several blogs about AutoCAD and AutoCAD VBA.

www.PacktPub.com

Support files, eBooks, discount offers and more

You might want to visit www.PacktPub.com for support files and downloads related to your book.

Did you know that Packt offers eBook versions of every book published, with PDF and ePub files available? You can upgrade to the eBook version at www.PacktPub.com and as a print book customer, you are entitled to a discount on the eBook copy. Get in touch with us at service@packtpub.com for more details.

At www.PacktPub.com, you can also read a collection of free technical articles, sign up for a range of free newsletters and receive exclusive discounts and offers on Packt books and eBooks.

PACKTLIB©

http://PacktLib.PacktPub.com

Do you need instant solutions to your IT questions? PacktLib is Packt's online digital book library. Here, you can access, read and search across Packt's entire library of books.

Why Subscribe?

- Fully searchable across every book published by Packt
- Copy and paste, print and bookmark content
- On demand and accessible via web browser

Free Access for Packt account holders

If you have an account with Packt at www.PacktPub.com, you can use this to access PacktLib today and view nine entirely free books. Simply use your login credentials for immediate access.

Table of Contents

Preface

Autodesk AutoCAD is, by far, the most used CAD software all around the world. In 2D it includes a large number of commands and functions, which makes it virtually unbeatable for many years. With continuous hardware development, together with new software capabilities, 3D became a viable and accessible technology to all. It is much better to simulate the real world with 3D models than to apply 2D drawings.

3D modeling has countless advantages: real-world simulation, greater accuracy, cheaper models, easy creation of related 2D drawings, calculation of volume and other properties, detection of interferences, model transfer to 3D printers or CAM/CNC devices, realistic visualization with light and materials application, sun studies, material selection, and easier comparison between solutions.

With Version 2007, AutoCAD became a reliable software for 3D modeling in all technical areas, such as architecture, engineering, construction, roads, urban studies, landscaping, and scenarios. Rendering and realistic results were quite improved in Version 2011, with the introduction of Autodesk Materials and the inclusion of more than 700 excellent-looking and ready-to-apply materials, as demonstrated on the cover of this book.

Autodesk AutoCAD 2013 3D Drafting and Design allows you to break the 2D frontier and create accurate 3D models that simulate reality. With the addition of lights and materials, simulation is taken to the level of photorealism. Including several explained exercises, this book is an easy learning tool and also a reference manual for daily consultation.

What this book covers

Chapter 1, *Introduction to 3D Design*, includes an introduction to 3D and the importance of the third coordinate. Also included are AutoCAD environment control, general object properties, auxiliary tools, 2D commands application, 3D linear commands, and good practices when modeling in 3D.

Chapter 2, Visualizing 3D Models, walks us through a fundamental aspect in 3D, which is visualization. This chapter includes not only zooming, panning, and orbiting, but also returning to specific visualizations, seeing the model as wireframe, shaded, or other visual styles, walking inside the model and dividing the drawing area into viewports.

Chapter 3, Coordinate Systems, includes the creation of other working planes called coordinate systems, imperative for correct 3D modeling.

Chapter 4, Creating Solids and Surfaces from 2D, walks us through the commands that allow for the creation of solids or surfaces from linear or planar objects.

Chapter 5, 3D Primitives and Conversions, includes the remaining commands to create 3D solids and 3D surfaces without previous objects and conversion commands.

Chapter 6, Editing in 3D, acquaints us with the editing commands specific for 3D operations that can be applied to any object.

Chapter 7, Editing Solids and Surfaces, presents all the main commands for combining solids and editing solids and surfaces. Among other useful commands, we can unite, subtract, intersect, and cut 3D objects, as well as apply fillets and chamfers to the object's edges.

Chapter 8, Inquiring the 3D Model, explains how to measure distances and volumes and obtain point coordinates. In 3D it is also important to detect interferences and obtain geometric properties of solids.

Chapter 9, Documenting a 3D Mode, provides guidelines for creating construction or fabrication drawings after frequently creating 3D models. In this chapter we present how to define a layout and then the most important commands and procedures to obtain automatic 2D drawings from 3D models.

Chapter 10, Rendering and Illumination, discusses that after creating a 3D model, it is time to present it as a virtual prototype or how it will look when built or fabricated. In this chapter we present the rendering process and all related commands, as well as simulating natural and artificial lighting.

Chapter 11, Materials and Effects, completes the render subject with materials and effects. As important as lighting a 3D scene, this chapter walks us through the application of realistic materials that resemble materials of the real world. AutoCAD also allows specifying scene backgrounds and applying fog effect.

Chapter 12, Meshes and Surfaces, is about all types of surfaces and meshes, including procedural surfaces, NURBS surfaces, meshes and polyface meshes.

Appendix, Final Considerations, includes creating simple animations representing walkthroughs or see-around, import and export file formats, advices for exporting from AutoCAD to 3ds Max and Revit, and development clues for 3D modelers.

What you need for this book

To correctly follow this book and realize all exercises, we need to have AutoCAD software, preferably the last version (2013 or later). Most of the book is also useful for other AutoCAD users since Version 2007; further improvements are specified along the book. Readers must also download exercise files from the book's webpage.

Who this book is for

This book is intended for everyone who wants to create accurate 3D models in AutoCAD, such as architects, engineers, or design professionals and students. Only some basic understanding of 2D AutoCAD is needed.

Conventions

In this book, you will find a number of styles of text that distinguish between different kinds of information. Here are some examples of these styles, and an explanation of their meaning.

Code words in text are shown as follows: "We may add this list to the Quick Access Toolbar by applying the CUI command or by right-clicking above the command icon we want to add."

A block of code is set as follows:

```
Command: CIRCLE
Specify center point for circle or [3P/2P/Ttr (tan tan radius)]: .X
of midpoint of edge
(need YZ): .Y
of midpoint of edge
(need Z): any point on top face
Specify radius of circle or [Diameter]: value
```

New terms and **important words** are shown in bold. Words that you see on the screen, in menus or dialog boxes for example, appear in the text like this: "To change it, we only have to click on the **OSNAP** button or press *F3*".

> [Warnings or important notes appear in a box like this.]

> [Tips and tricks appear like this.]

Reader feedback

Feedback from our readers is always welcome. Let us know what you think about this book — what you liked or may have disliked. Reader feedback is important for us to develop titles that you really get the most out of.

To send us general feedback, simply send an e-mail to feedback@packtpub.com, and mention the book title via the subject of your message.

If there is a topic that you have expertise in and you are interested in either writing or contributing to a book, see our author guide on www.packtpub.com/authors.

Customer support

Now that you are the proud owner of a Packt book, we have a number of things to help you to get the most from your purchase.

Downloading the example code

You can download the example code files for all Packt books you have purchased from your account at http://www.packtpub.com. If you purchased this book elsewhere, you can visit http://www.packtpub.com/support and register to have the files e-mailed directly to you.

Errata

Although we have taken every care to ensure the accuracy of our content, mistakes do happen. If you find a mistake in one of our books—maybe a mistake in the text or the code—we would be grateful if you would report this to us. By doing so, you can save other readers from frustration and help us improve subsequent versions of this book. If you find any errata, please report them by visiting http://www.packtpub.com/submit-errata, selecting your book, clicking on the **errata submission form** link, and entering the details of your errata. Once your errata are verified, your submission will be accepted and the errata will be uploaded on our website, or added to any list of existing errata, under the Errata section of that title. Any existing errata can be viewed by selecting your title from http://www.packtpub.com/support.

Piracy

Piracy of copyright material on the Internet is an ongoing problem across all media. At Packt, we take the protection of our copyright and licenses very seriously. If you come across any illegal copies of our works, in any form, on the Internet, please provide us with the location address or website name immediately so that we can pursue a remedy.

Please contact us at copyright@packtpub.com with a link to the suspected pirated material.

We appreciate your help in protecting our authors, and our ability to bring you valuable content.

Questions

You can contact us at questions@packtpub.com if you are having a problem with any aspect of the book, and we will do our best to address it.

1
Introduction to 3D Design

Welcome to the fantastic world of 3D! AutoCAD is an excellent software for creating 2D projects in all technical areas, but instead of 2D representation, isn't it much better if we could create accurate 3D models, view them from all perspectives (even from inside), and get 2D drawings easily? With AutoCAD we can!

The topics covered in this chapter are as follows:

- The importance of the third coordinate
- How to choose and manage 3D workspaces
- Why object properties are fundamental in 3D
- How auxiliary tools (osnap, ortho, and others) can ease the work in 3D
- 2D commands in a 3D world
- How to use linear 3D commands
- How to create great 3D models

The Z coordinate

3D is all about the third Z coordinate. In 2D, we only care for the X and Y axes, but never used the Z axis. And most of the time, we don't even use coordinates, just the top-twenty AutoCAD commands, the Ortho tool, and so on. But in 3D, the correct use of coordinates can substantially accelerate our work. We will first briefly cover how to introduce points by coordinates and how to extrapolate to the third dimension.

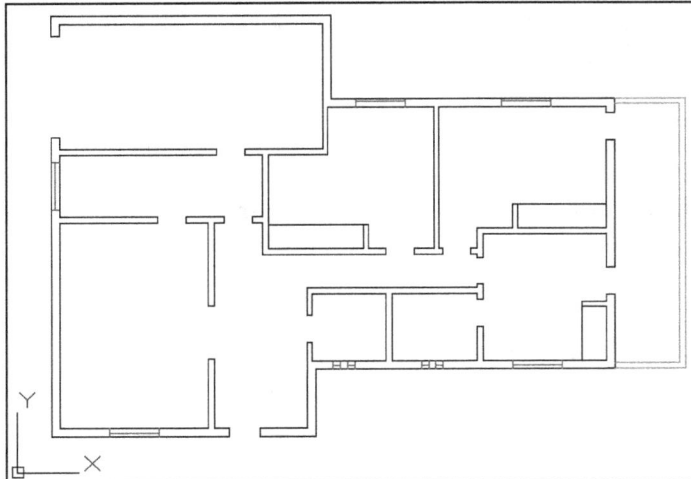

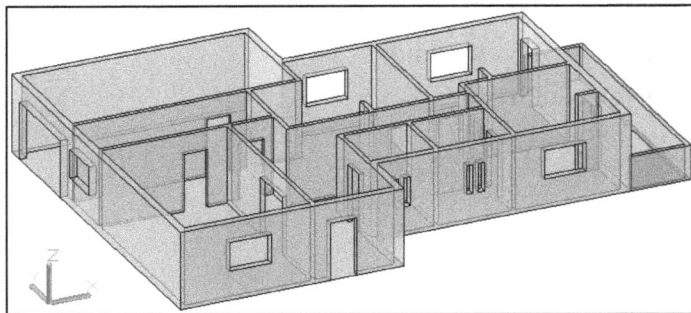

Absolute coordinates

The location of all entities in AutoCAD is related to a coordinate system. Any coordinate system is characterized by an origin and positive directions for the X and Y axes. The Z axis is obtained directly from the X and Y axes by the right-hand rule: if we rotate the right hand from the X axis to the Y axis, the thumb indicates the positive Z direction.

Picture that when prompting for a point; besides specifying it in the drawing area with a pointing device such as a mouse, we can enter coordinates using the keyboard.

The format for the absolute Cartesian coordinates related to the origin is defined by the values of the three orthogonal coordinates, namely, X, Y, and Z, separated by commas:

```
X coordinate, Y coordinate, Z coordinate
```

The Z coordinate can be omitted.

For instance, if we define a point with the absolute coordinates 30, 20, and 10, this means 30 absolute is in the X direction, 20 is in the Y direction, and 10 is in the Z direction.

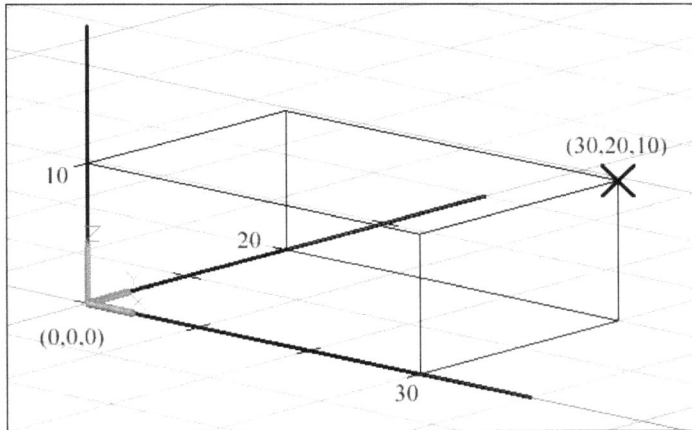

Relative coordinates

Frequently, we want to specify a point in the coordinates, but one that is related to the previous point. The format for the relative Cartesian coordinates is defined by the symbol **AT** (@), followed by increment values in the three directions, separated by commas:

```
@X increment, Y increment, Z increment
```

Of course, one or more increments can be 0. The Z increment can be omitted.

For instance, if we define a point with relative coordinates, @0,20,10, this means in relation to the previous point, 0 is in X, 20 is in Y, and 10 is in Z directions.

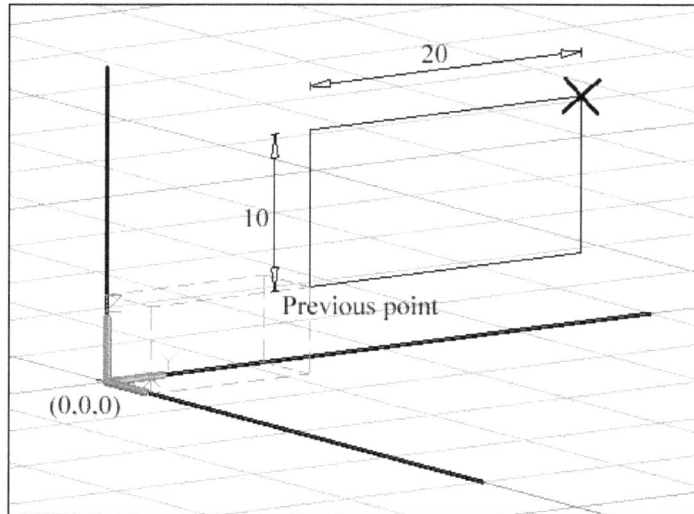

Point filters

When we want to specify a point but decompose it step-by-step, that is, separate its coordinates based on different locations, we may use filters. When prompting for a point, we access filters by digitizing the X, Y, or Z axes for individual coordinates, or XY, YZ, or ZX for pairs of coordinates. Another way is from the osnap menu, *CTRL* + mouse right-click, and then **Point Filters**. AutoCAD requests for the remaining coordinates until the completion of point definition.

Imagine that we want to specify a point, for instance, the center of a circle, where its X coordinate is given by the midpoint of an edge, its y coordinate is the midpoint of another edge, and finally its Z coordinate is any point on a top face. Assuming that Midpoint osnap is predefined, the dialog should be:

```
Command: CIRCLE
Specify center point for circle or [3P/2P/Ttr (tan tan radius)]: .X
of midpoint of edge
(need YZ): .Y
of midpoint of edge
(need Z): any point on top face
Specify radius of circle or [Diameter]: value
```

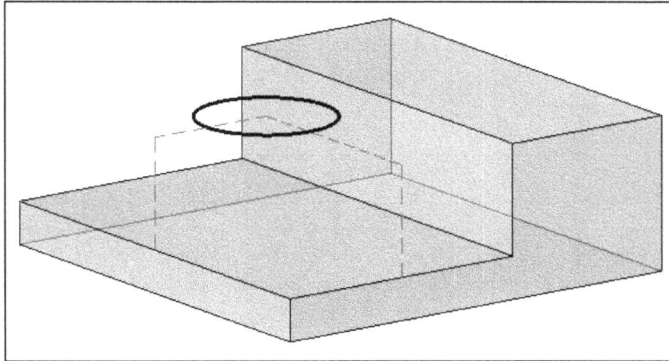

Workspaces

AutoCAD comes with several workspaces. It's up to each of us to choose a workspace based on a classic environment or the ribbon. To change workspaces, we can pick the workspace switching button on the status bar:

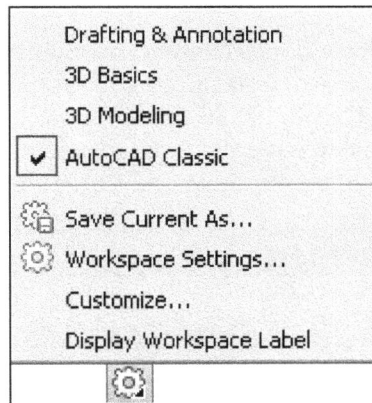

There are other processes for acceding this command such as the workspaces list on the Quick Access Toolbar (title bar), the Workspaces toolbar, or by digitizing WSCURRENT, but the access shown is consistent among all versions and always available.

Classic environment

The classic environment is based on the toolbars and the menu bar and doesn't use the ribbon. AutoCAD comes with AutoCAD Classic workspace, but it's very simple to adapt and view the suitable toolbars for 3D.

The advantages of using this environment are speed and consistency. To show another toolbar, we right-click over any toolbar and choose it. Typically, we want to have the following toolbars visible besides Standard and Layers: Layers II, Modeling, Solid Editing, and Render:

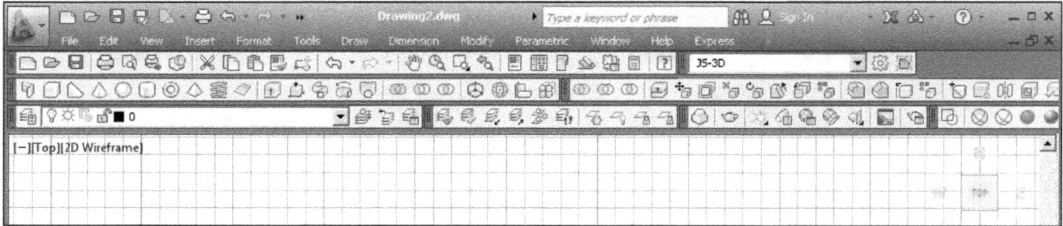

Ribbon environment

Since the 2009 version, AutoCAD also allows for a ribbon-based environment. Normally, this environment uses neither toolbars nor the menu bar. AutoCAD comes with two ribbon workspaces, namely, 3D Basics and 3D Modeling; the first being less useful than the second.

The advantages are that we have consistency with other software, commands are divided into panels and tabs, the ribbon can be collapsed to a single line, and it includes some commands not available on the toolbars. The disadvantage is that as it's a dynamic environment, we frequently have to activate other panels to access commands and some important commands and functions are not always visible:

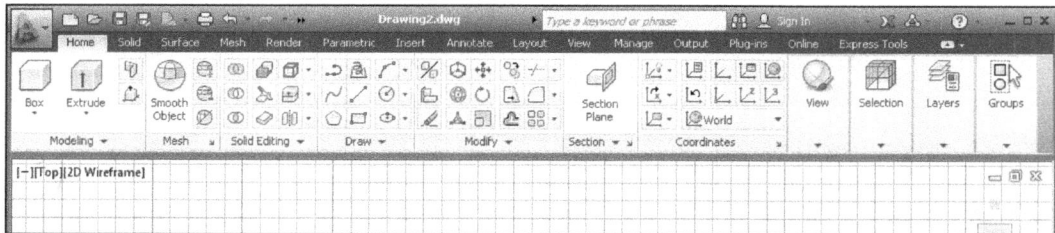

> When modeling in 3D, the layers list visibility is almost mandatory. We may add this list to the Quick Access Toolbar by applying the CUI command or by right-clicking above the command icon we want to add. Another way is to pull the **Layers** panel to the drawing area, thus making it permanently visible.

Layers, transparency, and other properties

When we are modeling in AutoCAD, the ability to control object properties is essential. After some hours spent on a new 3D model, we can have hundreds of objects that overlap and obscure the model's visibility. Here are the most important properties.

Layers

If a correct layers application is fundamental in 2D, in 3D it assumes extreme importance. Each type of 3D object should be in a proper layer, thus allowing us to control its properties:

- **Name**: A good piece of advice is to not mix 2D with 3D objects in the same layers. So, layers for 3D objects must be easily identified, for instance, by adding a 3D prefix.

- **Freeze/Thaw**: In 3D, the density of screen information can be huge. So freezing and unfreezing layers is a permanent process. It's better to freeze the layers than to turn off because objects on frozen layers are not processed (for instance, regenerating or counting for ZOOM **Extents**), thus accelerating the 3D process.

- **Lock/Unlock**: It's quite annoying to notice that at an advanced phase of our project, our walls moved and caused several errors. If we need that information visible, the best way to avoid these errors is to lock layers.

- **Color**: A good and logical color palette assigned to our layers can improve our understanding while modeling.

- **Transparency**: If we want to see through walls or other objects at the creation process, we may give a value between 0 and 90 percent to the layers transparency.

Last but not least, the best and the easiest process to assign rendering materials to objects is by layer, so another good point is to apply a correct and detailed layer scheme.

Transparency

Transparency, as a property for layers or for objects, has been available since Version 2011. Besides its utility for layers, it can also be applied directly to objects. For instance, we may have a layer called 3D-SLAB and just want to see through the upper slab. We can change the objects' transparency with PROPERTIES (*Ctrl + 1*).

To see transparencies in the drawing area, the **TPY** button (on the status bar) must be on.

Visibility

Another recent improvement in AutoCAD is the ability to hide or to isolate objects without changing layer properties.

We select the objects to hide or to isolate (all objects not selected are hidden) and right-click on them. On the cursor menu, we choose **Isolate** and then:

- **Isolate Objects**: All objects not selected are invisible, using the ISOLATEOBJECTS command

- **Hide Objects**: The selected objects are invisible, using the HIDEOBJECTS command

- **End Object Isolation**: All objects are turned on, using the UNISOLATEOBJECTS command.

There is a small lamp icon on the status bar, the second icon from the right. If the lamp is red, it means that there are hidden objects; if it is yellow, all objects are visible:

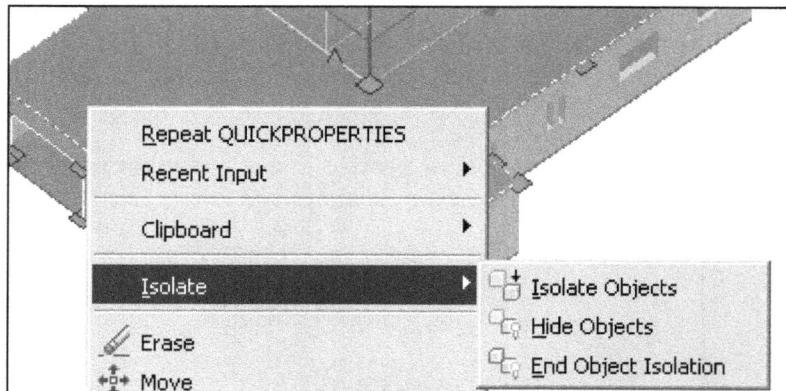

Shown on the following image is the application of transparency and hide objects to the left wall and the upper slab:

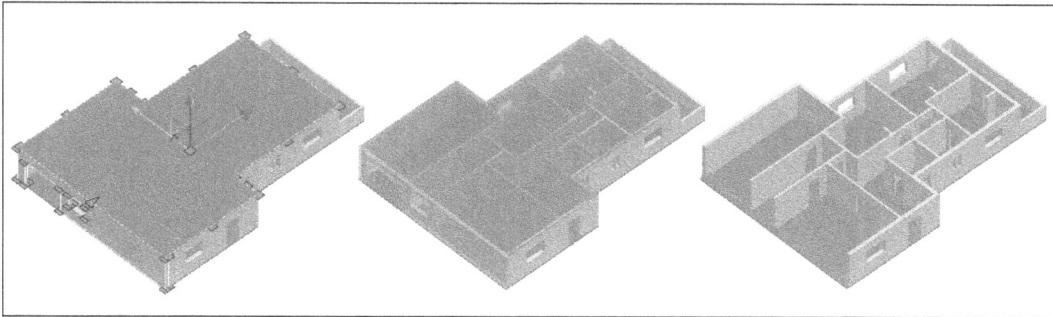

Auxiliary tools

AutoCAD software is very precise and the correct application of these auxiliary tools is a key factor for good projects. All users should be familiar with at least Ortho and Osnap tools. Following is the application of auxiliary tools in 3D projects complemented with the first exercise.

OSNAP, ORTHO, POLAR, and OTRACK auxiliary tools

Let's start with object snapping, probably the most frequently used tool for precision. Every time AutoCAD prompts for a point, we can access predefined object snaps (also known as osnaps) if the **OSNAP** button on the status bar is on. To change it, we only have to click on the **OSNAP** button or press *F3*. If we want an individual osnap, we can, among other ways, digitize the first three letters (for instance, MID for midpoint) or use the osnap menu (*CTRL* + right-click). Osnaps work everywhere in 3D (which is great) and is especially useful is the Extension osnap mode, which allows you to specify a point with a distance in the direction of any edge.

But what if we want to specify the projection of 3D points onto the working XY plane? Easy! If the OSNAPZ variable is set to 1, all specified points are projected onto the plane. This variable is not saved and 0 is assigned as the initial value.

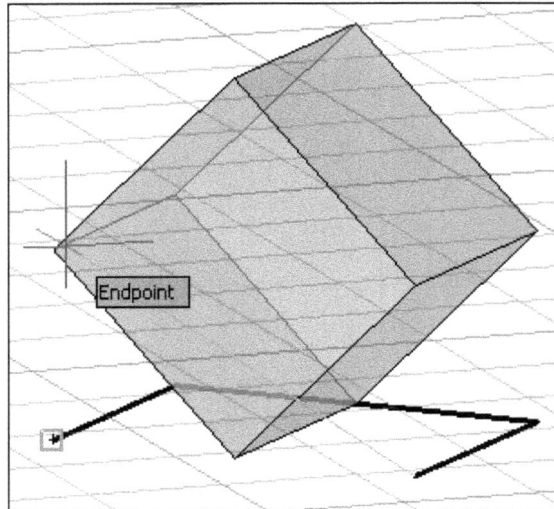

More great news is that **ORTHO** (*F8*) and **POLAR** (*F10*) work in 3D. That is, we can specify points by directing the cursor along the Z axis and assign distances. Lots of @ spared, no?

OTRACK (*F11*), used to derive points from predefined osnaps, also works along the Z-axis direction. We pause over an osnap and can assign a distance along a specific direction or just obtain a crossing:

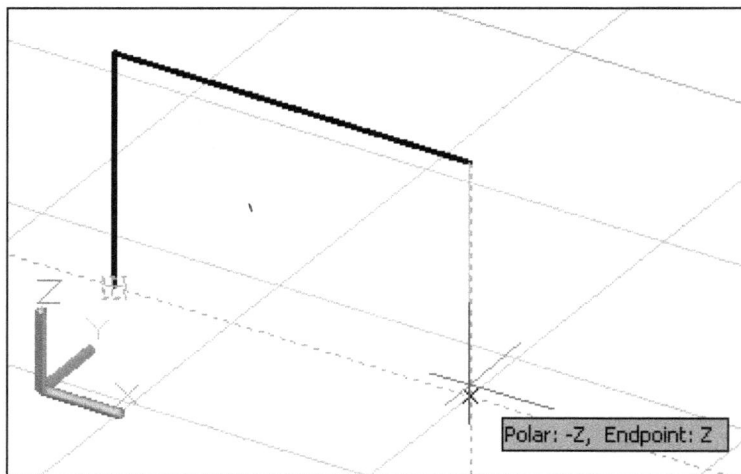

3DOsnap tool

Starting with Version 2011, AutoCAD allows you to specify 3D object snaps. Also, here we can access predefined 3D osnaps keeping **3DOSNAP** (*F4*) on, or we can access them individually. There are osnaps for vertices, midpoints on edges, centers of faces, knots (spline points), points perpendicular to faces, and points nearest to faces.

Exercise 1.1

Using the LINE command, coordinates, and auxiliary tools, let's create a cabinet skeleton. All dimensions are in meters and we start from the lower-left corner. The **ORTHO** or **POLAR** button must be on and the **OTRACK** and **OSNAP** buttons with Endpoint and Midpoint predefined.

> As in 2D, rotating the wheel mouse forward, we zoom in; rotating the wheel backward, we zoom out; all related to cursor position. To automatically orbit around the model, we hold down *SH!FT* and the wheel simultaneously. The cursor changes to two small ellipses and then we drag the mouse to orbit around the model. Visualization is the subject of the next chapter.

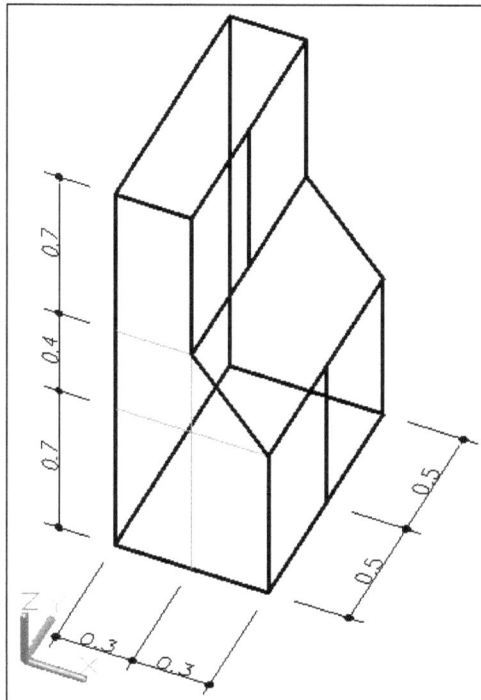

- We run the LINE command at any point, block direction X (**POLAR** or **ORTHO**) and assign the distance:

```
Command: LINE
Specify first point: any point
Specify next point or [Undo]: 0.6
```

- We block the Z direction and assign the distance:

```
Specify next point or [Undo]: 0.7
```

- The best way to specify this point is with relative coordinates:

```
Specify next point or [Close/Undo]: @-0.3,0,0.4
```

- We block the Z direction and assign the distance:

```
Specify next point or [Close/Undo]: 0.7
```

- The best way to close the left polygon is to pause over the first point, move the cursor up to find the crossing, with Polar or Ortho coming from the last point, and apply Close option to close the polygon:

```
Specify next point or [Close/Undo]: point with OTRACK
Specify next point or [Close/Undo]: C
```

- We copy all lines 1 meter in the Y direction:

```
Command: COPY
Select objects: Specify opposite corner: 6 found
Select objects: Enter
Current settings: Copy mode = Multiple
Specify base point or [Displacement/mOde] <Displacement>: point
Specify second point or [Array] <use first point as displacement>:
1
Specify second point or [Array/Exit/Undo] <Exit>: Enter
```

- We complete the cabinet skeleton by drawing lines between endpoints and midpoints:

```
Command: LINE
```

Application of 2D commands

Can those everyday commands be used in 3D? Of course they can! We have already seen the LINE command, layers, and other properties. Let's see some particularities and 3D applications and learn that a whole bunch of known commands can also be applied.

Drawing commands

Basically, all drawing commands can be used in 3D, provided that we have the correct working plane, LINE being the exception.

The LINE command can have its endpoints anywhere, so it's a real 3D command. But circles, arcs, and polylines (including polygons and rectangles) are drawn on the working plane (called active coordinate system) or a plane parallel to the working plane.

Editing commands

Here is the list of the most important editing commands that work the same way in 2D or 3D: ERASE, MOVE, COPY, SCALE, JOIN, EXPLODE, and BREAK.

Some commands work only on the objects plane, not necessary the active working plane. Examples are FILLET, CHAMFER, and OFFSET.

There are also some that work only in relation to the active working plane such as MIRROR, ARRAYCLASSIC (ARRAY before version 2012), and ROTATE.

Next are editing commands that have special 3D features:

- TRIM and EXTEND: Both commands have an option, **Project**, which specifies if linear objects are cut or extended to the boundaries related to the current coordinate system or the current view. This allows for cutting or extending objects that, in the current view, seems to be on the same plane, but are really on different planes.

- ARRAY: This command changed a lot in version 2012. Now, we have three different commands for rectangular, polar, and path arrays. Among multiple options, there are two with special importance for 3D: **Rows**, where we can define a number of rows with a height distance, and **Levels** also with a variation in height.

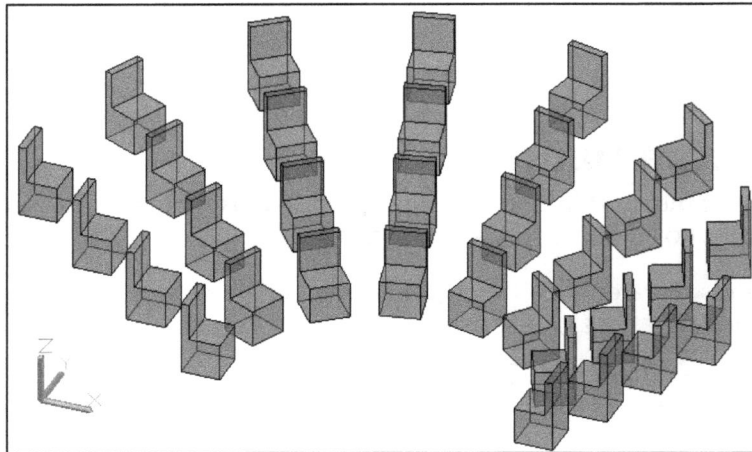

Other entities and commands

No one can use AutoCAD without inquiring for information from time-to-time. The DIST command allows you to obtain the 3D distance, and also increments in X, Y, and Z directions, between two points. Another important command is ID (or from the **Tools** menu bar, **Inquiry | ID Point**), for inquiring about the absolute coordinates of points.

Blocks work exactly the same way in 2D or 3D. When inserting a block with non-uniform scale, we can specify a different scale for the Z direction.

Regions are 2D opaque closed objects that are frequently used in 3D. Besides 3D, they can be very useful for extracting areas, inertia moments, and other geometric properties.

To create regions, we must have their contours already drawn. Contours can be lines, arcs, circles, ellipses, elliptical arcs, and splines. In 2D, regions are created with two commands:

- REGION: This command (alias REG) creates regions from closed objects or sets of linear objects that define a closed boundary. It only prompts for the selection of objects and the original objects by default are deleted.

- BOUNDARY: This great command (alias BO) allows for the easy creation of closed polylines or regions by specifying internal points to closed boundaries. An example follows.

Exercise 1.2

We are going to create some 2D objects and from them, some regions.

1. Start AutoCAD and create some 2D linear objects, similar to those shown in the following image. For now, dimensions are not important.

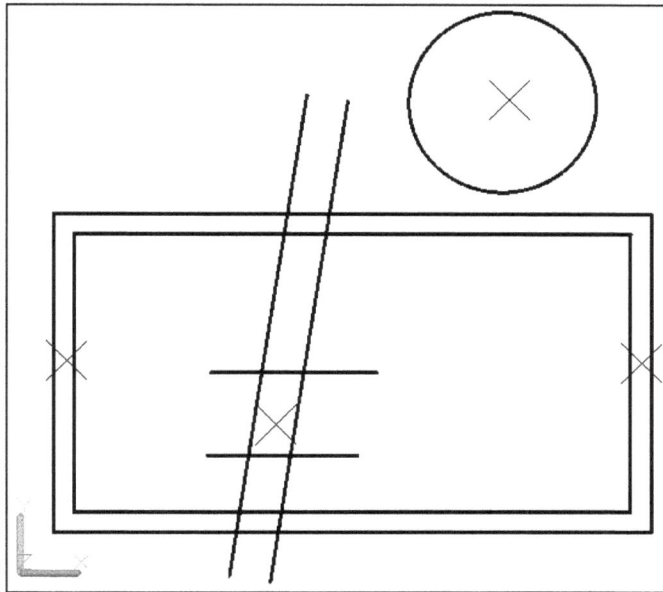

2. Create a new layer called REGIONS, assign a different color to it, and activate it.

3. Run the BOUNDARY command (or use BO alias). On the dialog box shown in **Object type** list, choose **Region**. Select the **Pick Points** button. The dialog box disappears.

4. Specify the four points inside the closed area, shown on the image and press *Enter*. The four regions are created. Freeze layer 0 to view only the regions.

5. Regions are opaque. To check, apply the VSCURRENT (VS) command (explained in the next chapter). Choose the **Realistic** option:

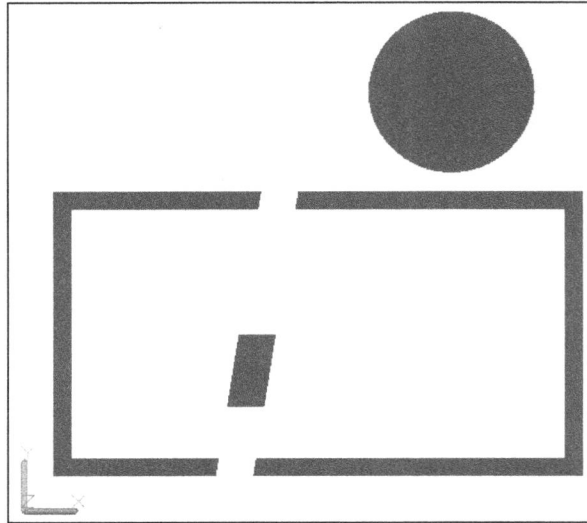

6. Apply the VSCURRENT command again and choose the **2dwireframe** option to come back to normal.
7. We don't need to save this drawing.

Linear 3D entities

We can apply thickness to most linear 2D entities and also create linear 3D entities like 3D polylines and splines.

Thickness

Almost all linear entities that we know from 2D have a property called Thickness, whose value represents a height along the Z axis (a better word actually should be height). This is the only 3D feature available in AutoCAD LT and can be applied to text (if made with a text style that uses an **SHX** font), but beyond that is quite limited.

The best way to change the value of thickness is with the PROPERTIES command, *Ctrl + 1*. A line is still a line or a text still text, but with a proper visualization, these entities can transmit a 3D feeling:

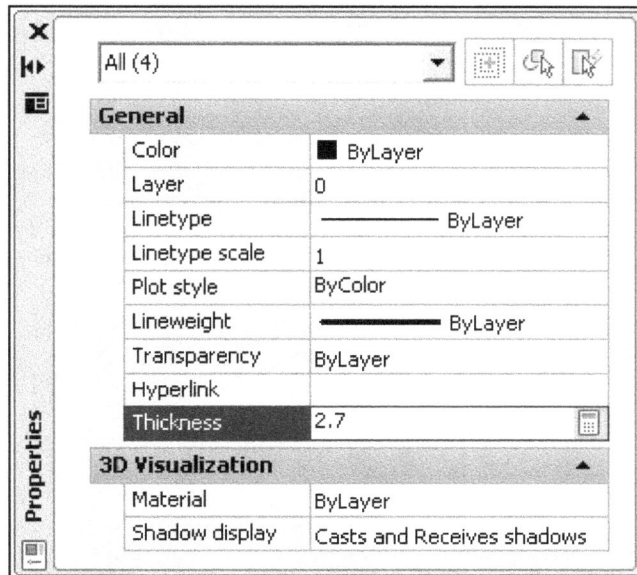

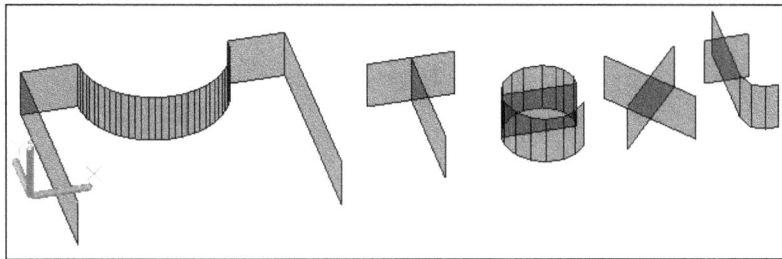

3D polylines

Polylines created with the PLINE command are 2D, not allowing for vertices with different Z coordinates. These polylines are designed by lightweight polylines. But what if a single object is needed to be composed by segments whose endpoints have different Z coordinates?

The answer is to create 3D polylines. Three processes are available:

- 3DPOLY: This command (alias 3P) creates 3D polylines from the start. It works like the LINE command, but the result is a single object.

- JOIN: This command (alias J) creates 3D polylines from a contiguous sequence of lines with shared endpoints. It is enough that one endpoint of a line is out of the plane for a 3D polyline to be created.

- PEDIT: The same command (alias PE) used for the creation of 2D polylines from lines and arcs can also be applied to join lines to a 3D polyline. The first object must already be a 3D polyline.

> Starting with version 2012, the best way to apply the JOIN command is to select all the line segments at the first command prompt, without specifying a source object. Depending on the type of selected objects and their positions, the most suitable object is automatically created.

And in what situations may we have utility for 3D polylines?

3D polylines are particularly useful to measure objects that develop in different directions, such as piping or wiring, and to define paths for other 3D objects such as piping. The creation of 3D solids and surfaces from linear objects is the subject of *Chapter 4, Creating Solids and Surfaces from 2D*.

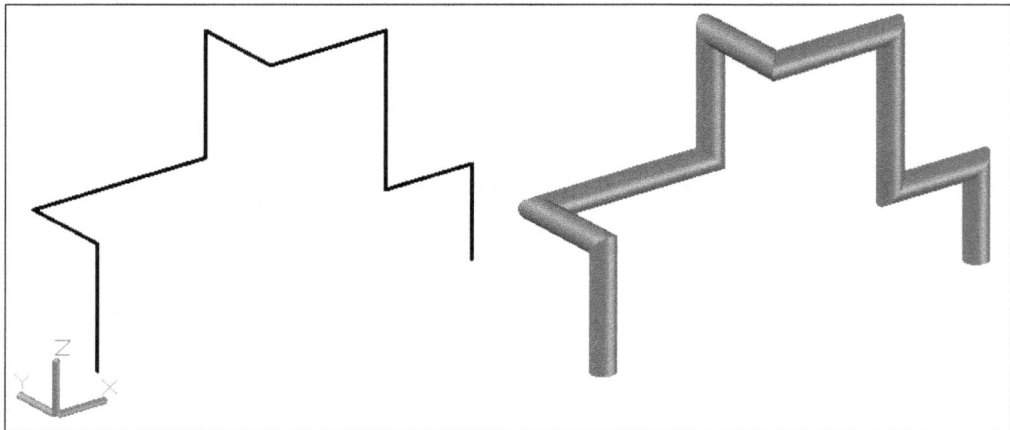

Splines and helixes

Splines are smooth linear objects, normally without corners that pass through or near specified points. Spline is short for **Nonuniform Rational B-Spline** (**NURBS**). Splines are described by a set of parametric mathematical equations, but have no fear, for AutoCAD will internally deal with this, we will not have to!

Splines are used whenever we need smooth curves and are also the foundation for NURBS surfaces, the most used surfaces in the automotive or aeronautic projects.

To create a spline we apply the (guess?) SPLINE command (alias SPL). By default, the command only prompts for the location of fit points, and the spline passes through the specified points. An *Enter* finishes the spline and the **Close** option creates a closed spline.

We can control splines by two sets of vertices:

- Fit points: These are the points used to create the spline
- Control vertices: These are the points that define the spline

When selected, the small blue triangle allows switching between Fit points and Control vertices. To edit a spline is very simple: we select it, without command, and work with grips. We can either click a grip and move it, or we can place the cursor over a grip and on the grip menu choose to move the vertex, add one, or remove that vertex. This last process is known as multifunctional grip.

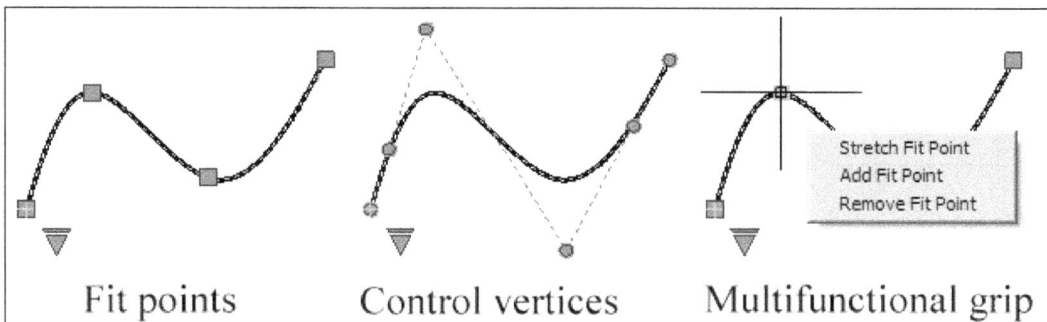

Fit points   Control vertices   Multifunctional grip

Since AutoCAD 2012, the JOIN command is also a great tool for the creation of complex splines. We may have a contiguous and non-planar sequence of lines, arcs, elliptical arcs, splines that the result is a single spline.

With the HELIX command we can create 2D or 3D helixes or spirals. By default, the command prompts for the center point of the base, base radius, top radius, and height. As options, we have the position of the axis endpoint, the number of turns, the height of one complete turn, and the twist (if the helix is drawn in the clockwise or the counterclockwise direction). To edit a helix object, we can use grips or the PROPERTIES command.

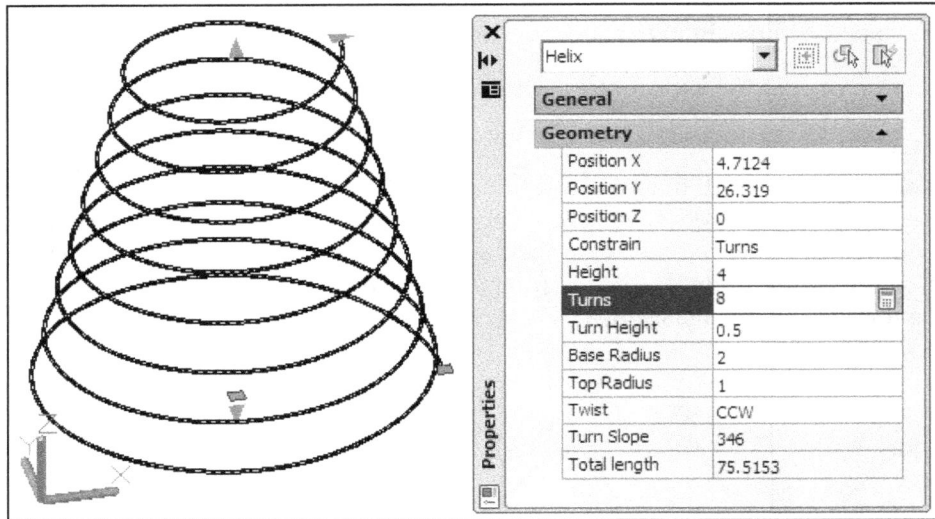

If we explode a helix object, the result is a spline. The BLEND command, new in Version 2012, creates a special spline that connects two open linear objects. The command only prompts for the selection of the first curve and the selection of the second curve. Selections must be near the endpoints to connect. The **Continuity** option allows for the choice of the applied type of continuity: **Tangent** with a tangency continuity (known as G1 continuity), or **Smooth** with a curvature continuity (known as G2).

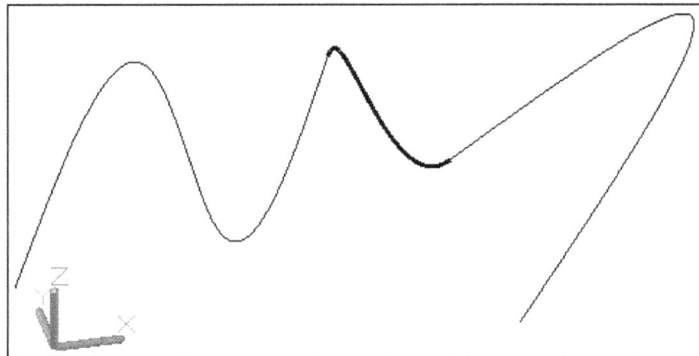

How to create great 3D models

There are several possibilities for starting a 3D project. We may:

- Create a 3D project from a complete set of 2D drawings, with all the necessary views included. Here, we don't have to project or idealize; we just have to decide the best approach and the commands that are to be used.

- Create a 3D project from a plant view and some other elements. Here we have to project a lot and probably have to study several possibilities in order to find the most suitable project.

- Create a 3D project from scratch. We have nothing except some conditions about space, functionality, or others.

For any of these possibilities, the keyword is planning.

First, we have to carefully plan the work. Instead of immediately starting to model, it's better and less time-consuming to decide a draft sequence of tasks.

Some important questions at this phase:

- Do I have access to something similar?
- How complex will my project be?
- Do I have some 3D blocks that I may use?
- Have I created the needed layers and other definitions (layouts, styles, and so on) in another project?
- Am I going to use external references in my project?
- Do I have all the necessary information?

If we have 2D drawings, these must be carefully studied, especially if there are any inconsistencies between views and how to start.

Next, we set up our model. We can open the most important 2D drawing and save it with a different name; we can start a drawing and insert the other drawings as blocks or external references. We must verify if units are coherent, by applying the UNITS command.

Continuing set up, we create layers and other definitions. A winning command here is ADCENTER, also known as Design Center (*Ctrl + 2*) that allows for gathering all layers, blocks, and other definitions from other drawings without opening them.

If we pretend to make some nice realistic images (rendering), we must be careful with layers, knowing that the easiest way to assign materials is by layer.

Should we start from floor plans or elevations? Well, it depends on the project. We can model from plans, but some parts may come from other views and then be positioned.

A final piece of advice is to keep several versions of your project. When an important step is achieved, we must save a backup copy. If we change our mind about a step or if the current file is corrupted, we minimize the losses.

Summary

In this chapter, we were introduced to 3D. We saw the importance of the third coordinate, how to enter points in absolute or relative coordinates, and the application of point filters. We analyzed workspaces and how to control the AutoCAD environment. The importance of layers, transparency, and other properties were then explained. Auxiliary tools were then covered before we looked at how to use them to ease the 3D project. We then saw the application of those commands in 3D used daily and some that are particularly important, such as BOUNDARY. We also covered the linear 3D commands and the Thickness property. We concluded the chapter by covering good practices when modeling in 3D.

2
Visualizing 3D Models

A fundamental aspect in 3D is visualization. This includes not only zooming, panning, and orbiting, but also returning to specific visualizations. We visualize the model as wireframe, shaded, or other visual styles, walk inside the model and divide the drawing area into viewports.

Topics covered in this chapter are as follows:

- How to view a 3D model from different angles
- How to save and restore particular views
- How to simulate cameras
- How to shade models and apply other visual styles
- How to navigate through a 3D model
- Dividing the drawing area into viewports

Zooming, panning, and orbiting

To correctly understand a 3D model, we need to view it from different angles, view it closer, from the inside, and so on. When creating or selecting objects, setting a correct and unambiguous view is fundamental. We are not scaling or rotating the model, just viewing it from a different perspective.

Zooming and panning

Zooming and panning are fundamental tools in AutoCAD and in 3D modeling. Nowadays, we all use a wheel mouse or other advanced input devices with scroll functionality. With a wheel mouse we can:

- **Zoom in / zoom out**: By rotating the wheel forward we zoom in and objects get closer; by rotating the wheel backward we zoom out and we see more of our scene.

- **Zoom extents**: By double-clicking on the wheel quickly, we see the drawing extents, that is, all the areas occupied by objects in the model get *extented* (including those areas that are visible when layers are turned off). This is quite useful in 3D and large models. Often, after orbiting, objects may disappear from the drawing area.

- **Pan**: By holding down the wheel and dragging the mouse, we slide the view, maintaining the same zoom factor.

Zooming and panning can be used transparently, in the middle of a command.

Zooming in/out with a wheel mouse is done by increments, not in a continuous way. Sometimes this is not acceptable for small zoom changes. The ZOOMFACTOR variable controls these increments. By default, it is set to 60%, that is, each wheel rotation increases or decreases the zoom factor by that value. The variable is saved in the registry. Another way for continuously changing the zoom factor is to use the ZOOM command (alias Z), the **Real-time** option. We enter the command ZOOM, press the *Enter* key to accept the default option, click and hold the mouse, and drag up or down to zoom in/out.

Orbiting the model

To view the model from different angles, we must orbit around it. The model doesn't change, only our angle of vision changes. We may use an automatic orbit or the commands presented next.

Automatic orbit

The easiest and most used tool to orbit a model is the automatic orbit with a wheel mouse. Even in the middle of a command, we hold down the *Shift* key and we simultaneously hold the wheel too; the cursor changes to two small ellipses, and then, we drag the mouse to orbit around the model. The Z axis is always vertical. It's as if we are flying around the model, as shown in the following screenshot:

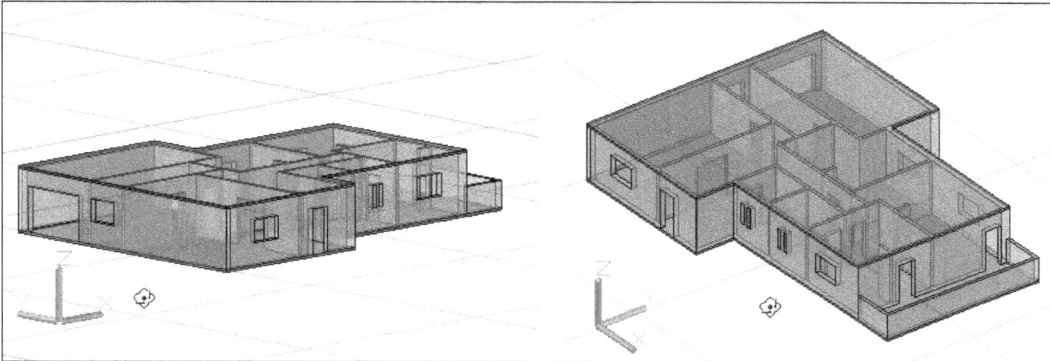

> While orbiting, only the objects that have been selected before are displayed. This can be particularly useful for checking their accuracy. When we end orbiting, all objects reappear.

The 3DORBIT command

The 3DORBIT command (alias 3DO), also known as **Constrained Orbit**, allows you to orbit by clicking and holding down the mouse. The Z axis is always vertical. If we precede its name or alias by a single quotation mark ('3DO) or if we access its icon on the **3D Navigate** toolbar, or by going to the **View | Navigate** panel on the ribbon, the command can also be used transparently (in the middle of another command).

Without a wheel mouse, this is the command to be used. The shortcut menu for this command (mouse right-click) gives direct access to several other commands and options, including additional commands for orbiting, zoom options, changing between parallel and perspective views, and predefined views and visual styles. Some of these commands and functions are presented next. The **Enable Orbit Auto Target** option must be highlighted as it allows for keeping the center of orbiting on the objects and not on the center of the drawing area:

Exit		
Current Mode: Constrained Orbit		
Other Navigation Modes ▶	✔ Constrained Orbit	1
	Free Orbit	2
✔ Enable Orbit Auto Target	Continuous Orbit	3
Animation Settings ...	Adjust Distance	4
Zoom Window	Swivel	5
Zoom Extents	Walk	6
Zoom Previous	Fly	7
✔ Parallel	Zoom	8
Perspective	Pan	9
Reset View		
Preset Views ▶		
Named Views		
Visual Styles ▶		
Visual Aids ▶		

> Also with this command, previously selected objects are isolated when orbiting. Ending the command restores all objects. With the **Zoom Extents** option, the isolated objects zoom to fit the drawing area, allowing for a better accuracy check.

Other orbiting commands

Other commands can be applied for orbiting around the model, and they have some features 3DFORBIT (**Free Orbit**): that may be useful:

- 3DFORBIT (**Free Orbit**): This is similar to the 3DORBIT command, but allows for a free orbit around the model, that is, it doesn't maintain the Z axis in the vertical direction. It shows a green circle, and the command behavior depends on the area where we drag. This command can be useful in the design, mechanical, or other areas where verticality isn't a key issue.

- `3DCORBIT` (**Continuous Orbit**): This allows for a continuous perpetual orbit. We drag the cursor in the required direction and the selected objects will visually rotate in that direction. A fast or slow drag will affect the orbit speed. The *Esc* key ends the command.

Exercise 2.1

Let's practice these visualization tools on a 3D model included in the accompanying files, as shown in the following steps:

1. Open the file `A3D_02_01.DWG`. This is the model shown in the previous image. Later we will create it in 2D.

2. Press the *Shift* key and the mouse wheel, and orbit around the model until you feel comfortable.

3. Without entering any command, select the left wall, which is separated from the other walls.

4. Press the *Shift* key and the mouse wheel, and orbit. Only the wall is being seen. As we end the dragging, all objects reappear. Press the *Esc* key to cancel the selection.

5. Without entering any command, select the balcony.

6. Type `3DO` to enter the `3DORBIT` command.

7. Right-click the mouse and choose **Zoom Extents** as shown in the following screenshot. The balcony is zoomed in to fit the drawing area:

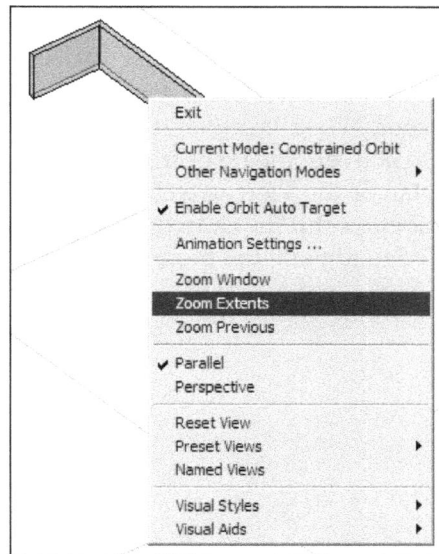

8. Click the mouse and orbit, seeing the balcony from all sides.

9. Right-click the mouse and choose the option **Other Navigation Mode | Free Orbit**. Click on the mouse and orbit, noticing that the Z axis is no longer vertical.

10. Again, right-click the mouse and choose the option **Other Navigation Mode | Constrained Orbit**. Drag a little bit and the Z axis comes immediately back to the vertical direction.

11. Press the *Esc* key to end the command. All other objects reappear, but the visualization doesn't change.

12. Quit the drawing without saving.

Views and cameras

If we changed the view direction, how can we come back to the plan view? Or see the model from another orthographic view? And how to save that view so that we can restore it anytime? These operations and the creation of cameras are presented next.

Predefined views

Predefined in AutoCAD, we have six orthographic views (**Top, Bottom, Front, Back, Left,** and **Right**) and four isometric views (**Southwest Isometric, Southeast Isometric, Northeast Isometric,** and **Northwest Isometric**). There are several ways to access them; some may activate another coordinate system.

View cube

The view cube, available since Version 2009, is a navigation tool placed by default in the viewport's top-right corner. When inactive, it is partially transparent. With this tool we can click on a face to activate the respective orthographic view. For instance, if we want to come to the plan or top view, we click the top face. We can also drag the mouse over the view cube to orbit the drawing, click a vertex to activate that isometric view, or click an edge to view the model from that edge.

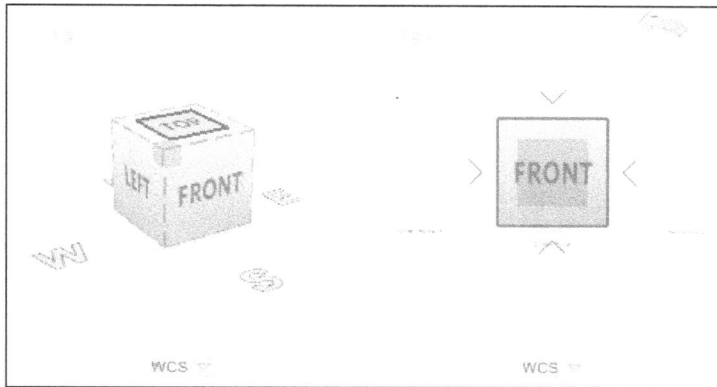

The view cube's properties are controlled by the shortcut menu (right-click over the cube) or the NAVVCUBE command. The small button below allows you to choose **world coordinate system (WCS)** or a user coordinate system. With the view cube, clicking on a face neither activates the corresponding coordinate system nor processes a ZOOM command's **Extents** option in the model.

Ribbon and toolbar access

The access to the predefined orthographic and isometric views can be found on the ribbon by going to the **Home | View** tab and clicking on the second drop list.

If we use a workspace based on toolbars, the **View** toolbar contains these predefined views.

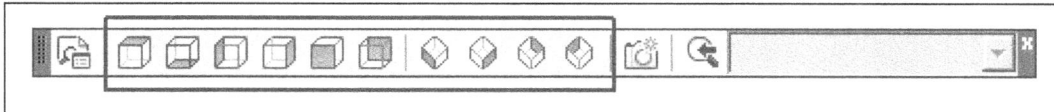

The ZOOM command's **Extents** option is processed in the scene when using these tools. Choosing an orthographic view, AutoCAD also activates the corresponding orthographic coordinate system, so the working XY plane becomes parallel to the view. This behavior may be undesirable and can be turned off by assigning the value 0 to the UCSORTHO variable. This variable is saved in the registry and doesn't depend on drawings.

The Viewport controls

Since Version 2012, in the top-left corner of the viewport, by default, there are three controls. The second one allows the activation of the predefined or saved views. Viewport controls are presented later.

The PLAN command

This command directly activates the **Top** view for the current coordinate system (default), any saved coordinate system, or the world coordinate system. It processes the ZOOM command's **Extents** option in the model.

The VPOINT and DDVPOINT commands

Two more commands are available for changing the viewing direction:

- VPOINT: This command (alias -VP) defines the viewing direction by a vector that connects a point to the origin. The command just asks for this point. For instance, looking at the origin, **0,0,1** indicates that we are on the Z axis; it means this is the top view. The ZOOM command's **Extents** option is always processed, so **0,0,2** will give exactly the same result. **1,1,1** means that we are at the same distance from all the positive axes (**Northeast Isometric**). This command has the advantage of replicating a precise non-orthographic view. The other options are less important.

- DDVPOINT: This command (alias VP) shows a dialogue box where the viewing direction is defined by two angles, the first on the plane from the X axis, and the second from the XY plane. We can enter these values or just pick on the images.

> Changing the view direction, such as activating an orthographic view or applying VPOINT generates a small animation. The same goes on when using the ZOOM command's **Extents** option, the **Window** option, or some other options. The VTOPTIONS command allows you to eliminate this animation. On the dialog box we uncheck the **Enable animation for pan & zoom** or **Enable animation when view rotates** option.

Creating views

The VIEW command (alias V) creates and manages saved views. On the ribbon, we can also access this command on the **View** panel inside the **View** tab:

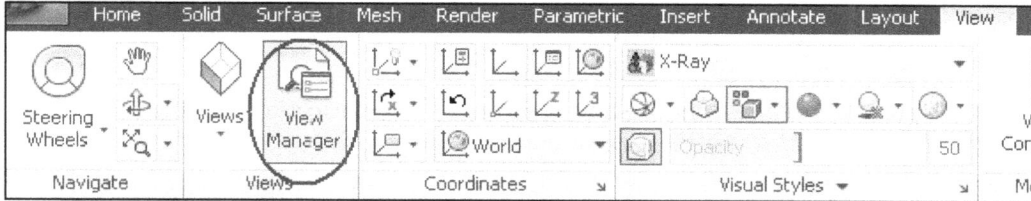

The command shows the **View Manager** box with three areas:

- **Views**: The area cn the left shows the available views that are divided by **Model Views**, **Layout Views**, and **Preset Views**. **Preset Views** includes the predefined orthographic and isometric views. After selecting a view, its properties are shcwn in the central area.

- **View**: The central area displays the properties of the selected view on the left.

- **Buttons**: The buttons on the right allow for activating a view, creating a new view, updating layers associated to a view, editing its boundaries, and deleting a view.

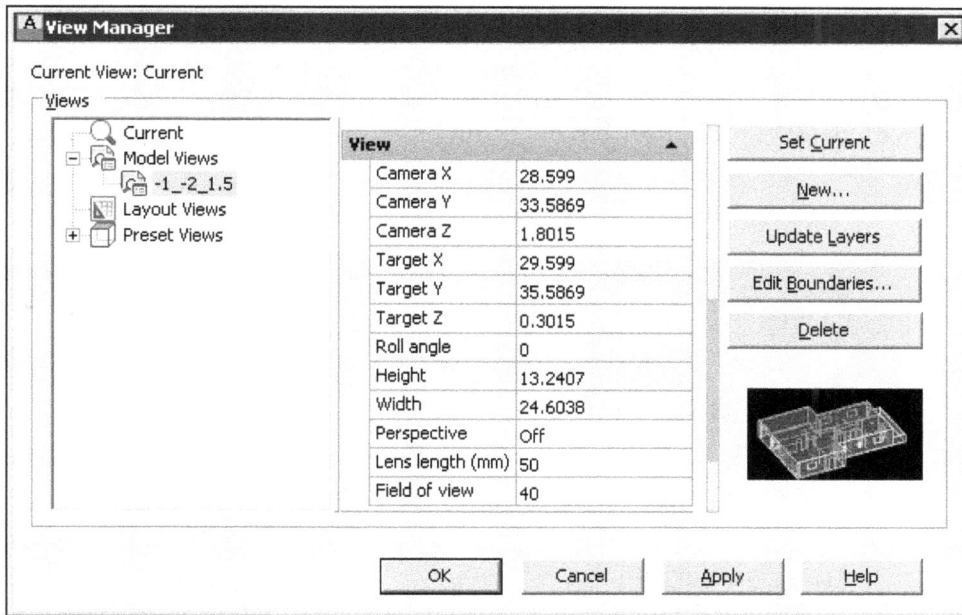

To create a view, we click on the **New** option. The **New View/Shot Properties** box is displayed. The most important elements are:

- **View name**: The name that identifies the view.
- **View category**: If there are too many views, these can be grouped in categories, for instance, exterior or interior views. We may maintain the default category **<None>**.
- **View type**: We can choose a **Still** view, a **Cinematic** view, or a **Recorded Walk** view. Normally we want to save a **Still** view.
- **View Properties**: There are three areas in this tab: The **Boundary** area specifies if the saved view is exactly the current display, or we may, by clicking the button, define a new window. The **Settings** area allows saving the layers with the view, coordinate system, live section (if any), and visual style are saved. We can also associate and configure a background to the view.

Associating the layers' properties to the saved view is very useful. By restoring a view, we will get the same layers' properties, such as frozen or thawed layers, when the view was created. The **Update Layers** button allows for assigning the current layer's properties to the selected view. The **Edit Boundaries** button allows for zooming in or out of the selected view. It displays an expanded drawing area with the current view having the window background.

The Perspective view

In the real world, we have a perspective projection. Objects farthest look smaller and parallel lines or edges look concurrent. When working in 3D models, it is easier to work with parallel projection, but for getting realistic final results we must turn perspective on:

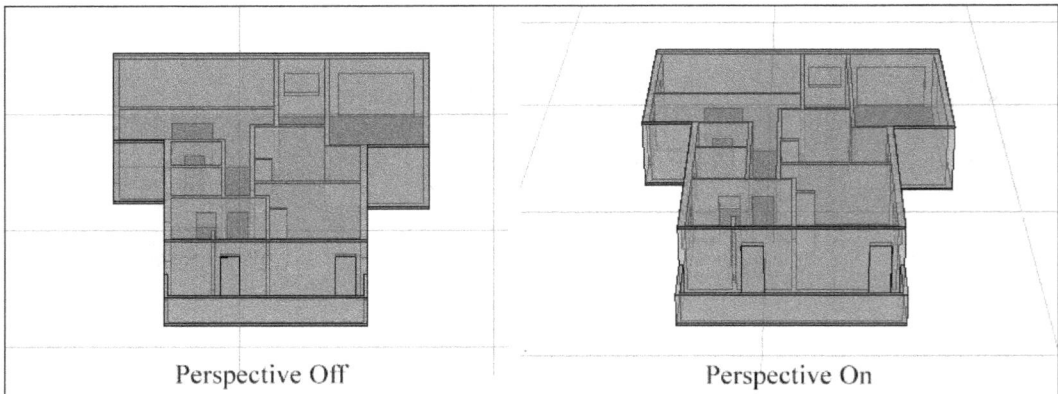

Perspective Off   Perspective On

To change between a parallel view and a perspective view, we can use the **PERSPECTIVE** variable: **1** to activate the perspective view and **0** to activate the parallel view. We can also use the **Parallel** and **Perspective** options on the shortcut menu over the view cube or the second viewport control.

Cameras

All 3D programs have the simulation of cameras, and AutoCAD is no exception. Cameras are very handy because they simulate a real camera with position, target, and lens length. Cameras can be moved and we can populate a 3D model with cameras. At any time we can activate a particular camera view.

The camera's creation

The CAMERA command (alias CAM) creates cameras. It prompts for the camera position and the target position, and displays a set of options. We press the *Enter* key; these options, if needed, can be changed later. Each camera is represented in the model by a wireframe blue object with a camera aspect, called camera glyph. The camera and target positions define the viewing direction. Cameras always have perspective projection. Camera views are always added to the **Model Views** group.

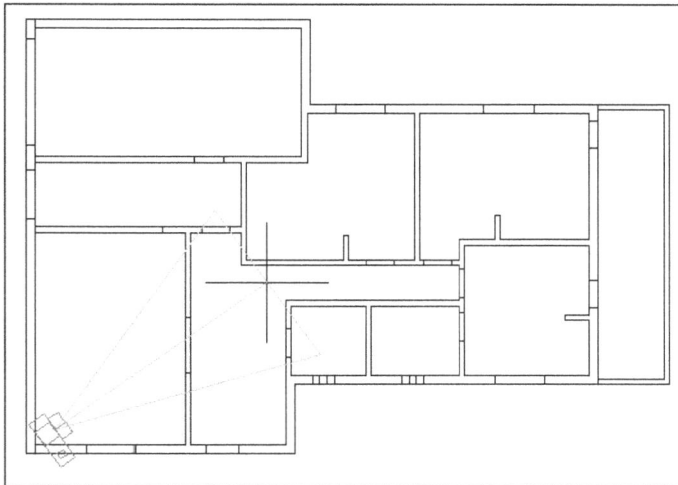

The camera's visualization and properties

When we select a camera, besides displaying its cone, a preview window is also displayed, allowing for a better control of the camera view. Options are available for choosing the visual style and checking if this preview window is displayed upon camera selection.

We can modify cameras in two ways:

- Using the PROPERTIES command, which is an excellent command. It allows for changing all objects' parameters and is also suitable for changing the camera's properties. We must activate this command's palette before selecting the camera. We can control the camera's name (**Name**), the camera's position coordinates (**Camera X**, **Camera Y**, and **Camera Z**), the target's position coordinates (**Target X**, **Target Y**, and **Target Z**), the camera's magnification or zoom (**Lens length (mm)** and **Field of view**), the camera's rotation around the viewing direction (**Roll angle**), and check if the camera representation object is plotted. Cameras can have visual clipping planes turned on and we can control the distances from the target position to the front clipping and back clipping planes (**Front plane** and **Back plane**).

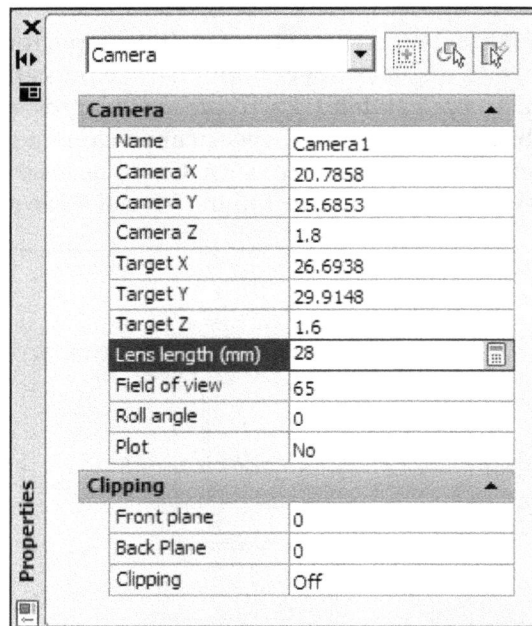

- Using grips is another way to modify the cameras. When we select the camera, a cone with grips is displayed and we can activate the grips to change the besides positions of the camera and target, pan the camera, and change the lens length.

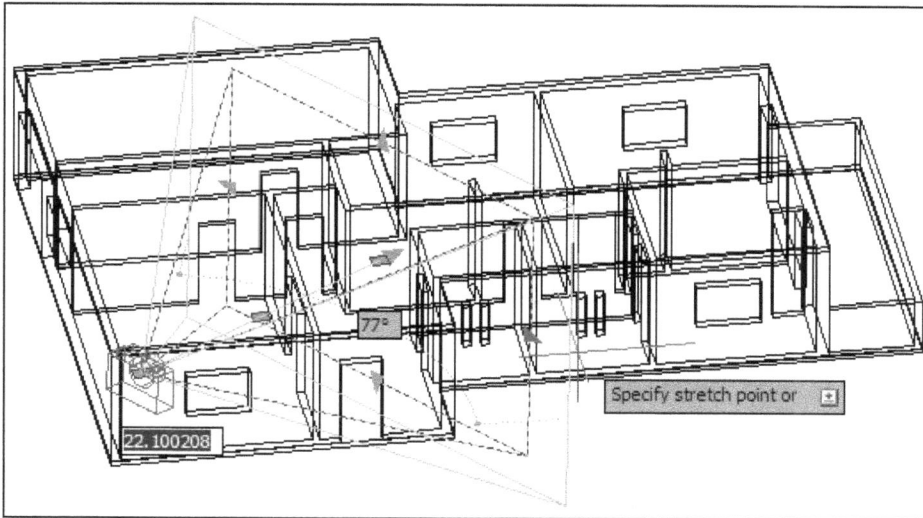

The camera's activation

To activate a camera view, as all cameras are added to Model Views, we can use any restore view available, such as the VIEW command, the second viewport control, the **View** toolbar, or the ribbon. Another way is to use the camera's shortcut menu and select **Set Camera View**.

> The best way to create and control a camera is to specify the camera's position and the target on the **Top** view, and then, use the PROPERTIES command to give a name, control the camera's height and the target's height (for instance, to simulate the height of someone looking at the model), and adjust the lens length.

Exercise 2.2

With the model used in *Exercise 2.1*, but with a conceptual visual style, we are going to create and apply views and cameras.

1. Open the file A3D_02_02.DWG.

2. Hide the top slab object. Select it and, on the shortcut menu, choose **Isolate/ Hide Objects**.

3. Using the view cube, select the **Front** view. Then select the **Right** view, the **Back** view, and the **Left** view. Each time we select an orthographic view, the view direction rotates to display the chosen view.

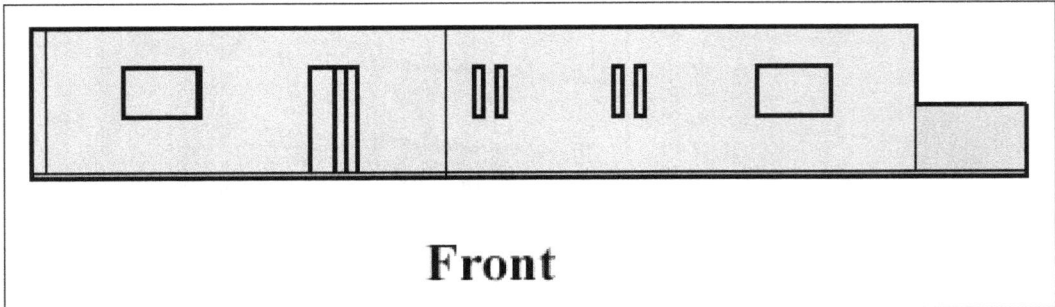

Front

4. We want a view that corresponds to a small rotation of the **Top** view, which when related to the model, will be positioned on the negative X and Y axis and the positive Z axis. Apply the VPOINT command and specify the value **-1,-2,1.5**. The view rotates, so we have a nice view.

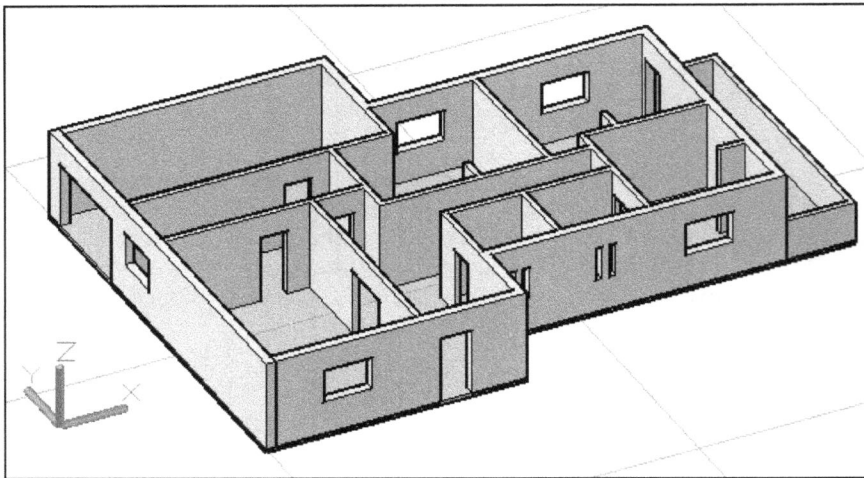

5. Save this view. Apply the VIEW command (alias V) and press the **New** button. In the **New View / Shot Properties** box, name the values **-1_-2_1.5** as name (view names don't accept commas, semicolons, or spaces). Don't change the remaining parameters. Press the **OK** button twice to end the command.

6. Select the perspective view. Place the cursor over the view cube, and with the shortcut menu, select **Perspective**. The background colors change to reflect the floor, sky, and the horizon line. Orbit the model to perceive these elements. These colors can be configured with the OPTIONS command (alias OP), the **Display** tab, and the **Colors** button.

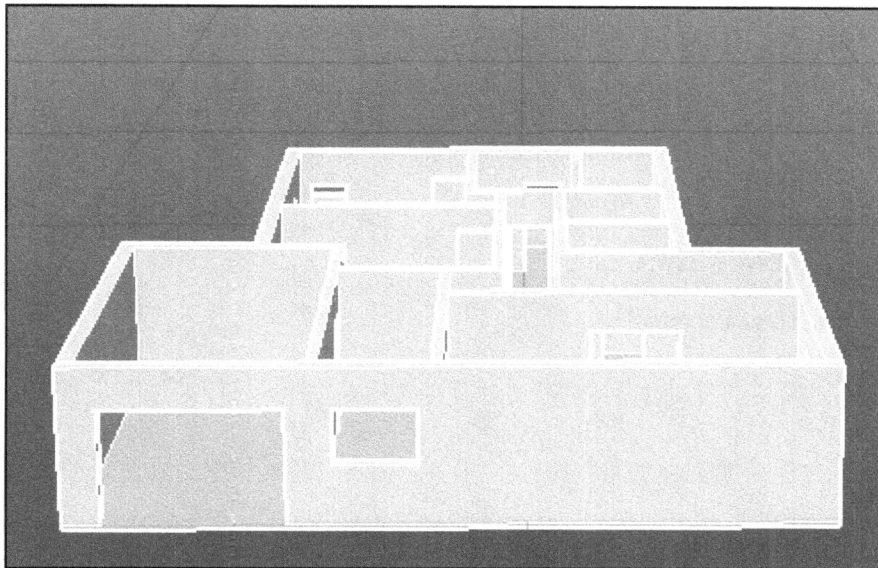

7. Undo the commands until you come back to the parallel view.

8. With the PLAN command, change to the **Top** view for the current UCS (actually the world coordinate system). A **Zoom Extents** option is automatically processed.

9. Apply the CAMERA command (alias CAM) to create a camera placed in the lower-left room and target the corridor. Press the *Enter* key to end the command.

10. Display the **Properties** palette (alias *CTRL + 1*). Select the camera. The **Camera Preview** window is displayed and the **Properties** palette reflects the camera's parameters. Change the following parameters: **Name** to Living room, **Camera Z** to 1.8, and **Target Z** to 1.7.

11. On the **Camera Preview** window, change the visual style to **Realistic**. The preview is not famous, so we need to open the field of view. On the **Properties** palette, change **Lens length (mm)** to `18`. Now, it is possible to see almost the entire living room.

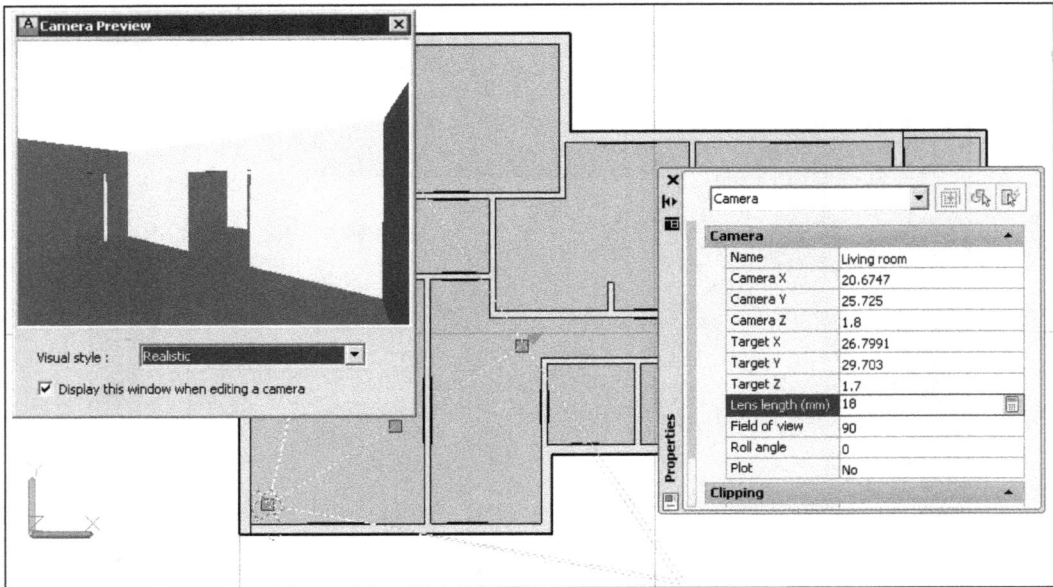

12. To activate the camera view, with the camera selected, on the shortcut menu specify **Set Camera View**.

13. Press the *Esc* key to end the selection and save the model with the name `A3D_02_02final.DWG`.

Visual styles

In the previous pages, we already have seen some visual styles applied. Visual styles, introduced in Version 2007, specify the representation of the 3D model on each viewport, such as controlling edges, shading or wireframing, lighting, materials, and background.

Applying visual styles

AutoCAD comes with several visual styles already defined that cannot be eliminated and are available for all drawings. Since Version 2011, there are ten visual styles available. Previous versions have only five.

- **2D wireframe**: This visual style only displays edges and is the only one that doesn't accept the perspective mode or background

- **Wireframe**: This visual style displays edges but accepts perspective mode and background

- **Hidden**: This visual style only displays visible edges, giving the feeling of opaque faces (hidden edges not visible)

- **Sketchy**: In this visual style, visible edges are displayed as if they were sketched by hand

- **Realistic**: In this visual style, faces are displayed with a smooth shade and with materials, if applied

- **Shaded**: In this visual style, faces are displayed with a smooth shade and without materials

- **Shaded with Edges**: This visual style is the same as shaded, but with visible edges highlighted

- **Shades of Gray**: This visual style is the same as shaded, but all colors are converted to grayscale

- **Conceptual**: In this visual style, faces are displayed with smooth shading, but with cool and warm colors, giving an artistic feeling

- **X-ray**: In this visual style, objects are displayed with some transparency

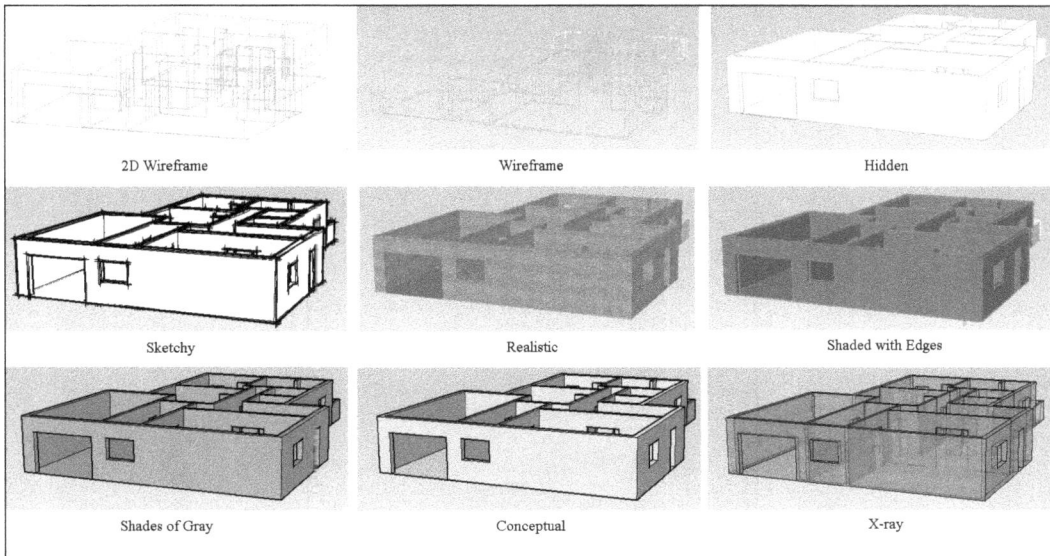

The VSCURRENT command (alias VS) specifies the visual style for the current viewport. It works from the command line, which means that it is the quickest way for setting a predefined visual style. For instance, to set the **Realistic** style, we press the sequence VS and press the *Enter* key, and then R followed by the *Enter* key. To set the 2D Wireframe style, we press VS and the *Enter* key, and then **2**, followed by the *Enter* key.

Other possible ways to choose another visual style are the third viewport control (since Version 2012),the **Visual Styles** panel in the View tab on the ribbon, the **Visual Styles** toolbar, or the 3DORBIT command shortcut menu.

Creating and modifying visual styles

Normally, the ten visual styles that come with AutoCAD are enough, but we may create others. The VISUALSTYLES command (alias VSM) displays a palette for creating and modifying visual styles.

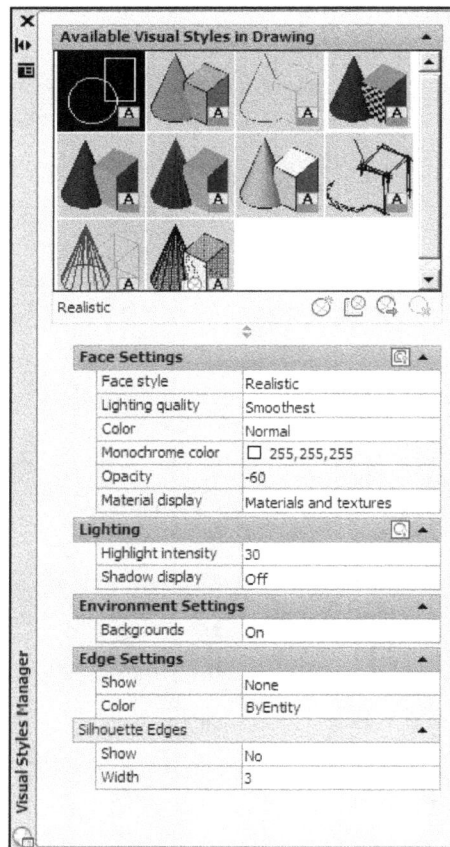

Below the **Available Visual Styles in Drawing** area are buttons to create a new visual style, to apply the selected visual style, to export the selected visual style to the **TOOL PALETTES** command palette, thus making it available to other drawings, and to delete the selected visual style. The shortcut menu over one of the slots also has options to apply it to all viewports, edit the description, copy and paste, modify the sample's size, and reset to the default parameters.

Visual styles' parameters are divided into:

- **Face Settings**: This parameter controls the face style (**Realistic, Gooch**, or **None**), lighting quality, color, opacity, and the material's display
- **Lighting**: This parameter controls the highlight intensity and the shadow's display
- **Environmental Settings**: This parameter controls the background
- **Edge Settings**: This parameter controls if the edges are displayed and how they are displayed, their color, and their sketchy effect

Exercise 2.3

Better than an exhaustive description of all parameters is to apply a practical example. Besides the specified values, we may play around with options and values.

1. Open the file A3D_02_03.DWG. The visual style of this model is **Realistic**, with the materials and shadows options turned on.

2. Apply the VSM alias to call **Visual Styles Manager**. We are going to create a new visual style, without materials and shadows, and that is partially transparent and yellow.

3. Click on the **Create New Visual Style** button (the first one) and name it My Yellow X-Ray.

4. Click on the second button to apply this visual style to the viewport so that any changes are immediately displayed.

5. In the **Face Settings** option, click the small button named **Opacity** to turn it on. In **Color** specify **Tint**, and on **Tint Color** choose a yellow color. Don't change the value of **Opacity** (it should be **60**), and **Material display** should already be turned off.

6. In the **Lighting** option, click on the small button named **Highlight intensity** and specify **40**. **Shadow display** should be off.

7. Inside the **Edge Settings** option, specify **Show** as **Isolines**, **Number of lines** as **0**, **Color** as **Blue**, and **Always on top** as **Yes**. And inside **Occluded Edges**, specify **Show** as **No**, and do the same with **Intersection Edges** and **Silhouette Edges**. In **Edge Modifiers**, the small buttons allow for the small lines' extension and sketchy mode (**Jitter**), but don't modify these.

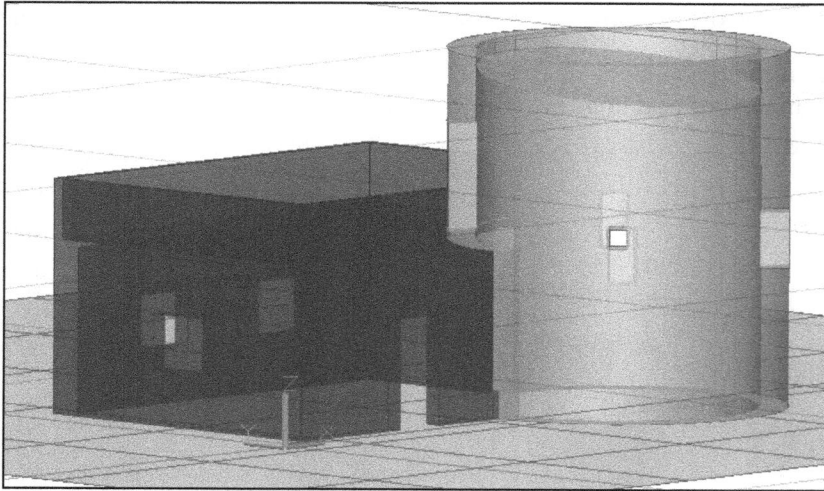

8. Click on the third button, **Export the Selected Visual Style to the Tool Palette**, and this visual style is placed on the default **TOOL PALETTES**. The correct way should be to first open **TOOL PALETTES** (*Ctrl + 3*), specify the **Visual Styles** palette, and then use the button.

9. Save the model with the name A3D_02_03final.DWG.

Walking and flying in a 3D model

For complex projects or intricate mechanisms, it may be useful to walk around or inside a model using both a keyboard and a mouse, such as a computer game. Besides walking (the Z coordinate constant), we can also fly (the Z coordinate variable).

Walking and flying

There are two commands for walking and flying. Both can be found on the **Walk and Fly** toolbar, the **View | Walk and Fly** menu bar, and ribbon's **Render | Animations** panel:

- 3DWALK: This command (alias 3DW) allows you to walk through or around the model while maintaining the Z coordinate

- 3DFLY: This command (no alias) allows you to fly through or around the model while changing the Z coordinate

These commands only work in the perspective mode. When entering one of these commands, if the model is in the parallel mode, the activation of the perspective mode is requested. Then, a small green cross, which is the target, is positioned at the center of the viewport and the **Position Locator** palette is displayed. This palette has two areas: on top we have a map scene and the user's position; below there are some view properties available, namely user height and target height.

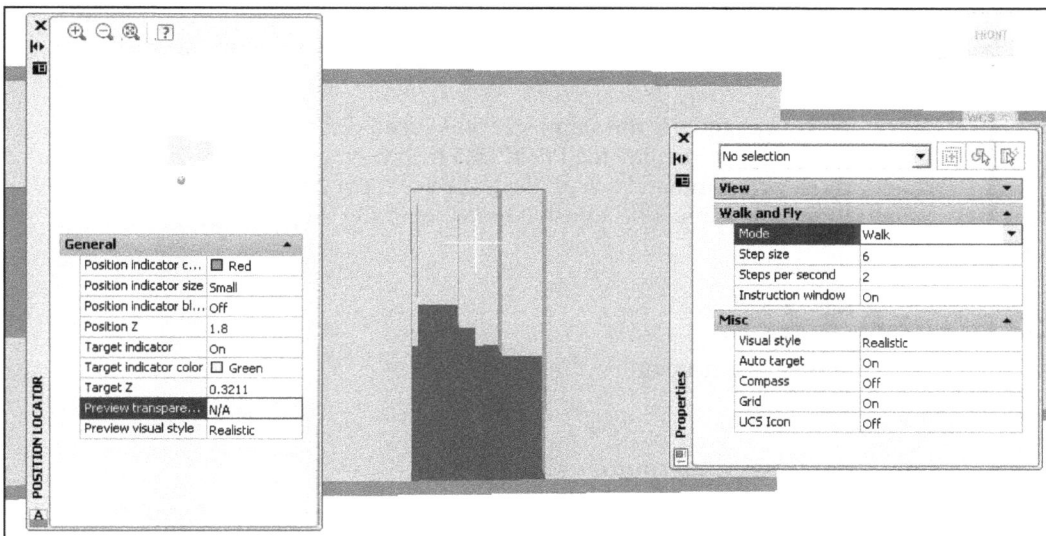

Instructions for both commands include the following:

- **Mouse**: By dragging the mouse, we control where to go. The view rotates so that the target is always at the center of the viewport
- *W* **key or up arrow**: By pressing the *W* key or the up arrow, we go forward
- *S* **key or down arrow**: By pressing the *S* key or the down arrow, we go backward
- *A* **key or left arrow**: By pressing the *A* key or the left arrow, we pan the view to the left
- *D* **or Right arrow**: By pressing the *D* key or the right arrow, we pan the view to the right
- *F* **key**: By pressing the *F* key, we change between walk and fly mode

Settings

When walking or flying, each time we press a movement key, we walk or fly a step. If we keep the key pressed, we move at speed. The step and speed parameters are easily controlled on the **PROPERTIES** palette (*Ctrl + 1*). We can modify the **Mode**, **Step size** (in drawing units), and **Steps per second** (speed) properties of walk/fly. Also, it is possible to change the visual style without ending the command.

A second possibility is the WALKFLYSETTINGS command. This one displays a box where we can specify the display of an instructions balloon, if the position locator will be displayed, **Walk/fly step size** in drawing units, and **Steps per second**.

> As we can modify the step's size and speed during walk or fly, a good tip is to display the **PROPERTIES** palette before entering these two commands. With the WALKFLYSETTINGS command, these values must be adjusted before entering the commands.

Exercise 2.4

It's time to walk inside our house.

1. Open the file A3D_02_04.DWG.
2. Apply the PROPERTIES command (*Ctrl + 1*).
3. Apply the command 3DW. On the **Walk and Fly - Change to Perspective View** box, click **Change** to confirm the perspective mode.
4. On the **PROPERTIES** palette, change **Lens length (mm)** to **28**, **Step size** to **1.5**, and maintain **Steps per second** with **2**.

5. On the **POSITION LOCATOR** palette, change **Position Z** to **1.8** and **Target Z** to **1.4**.

6. Use keyboard arrows to walk in the house and drag the mouse key to maintain the target aligned with the opening of the door.

7. Once inside, drag the mouse to see around. Continue walking inside the house.

8. While using **3DWALK**, we can change the visual styles, lens length, and other parameters:

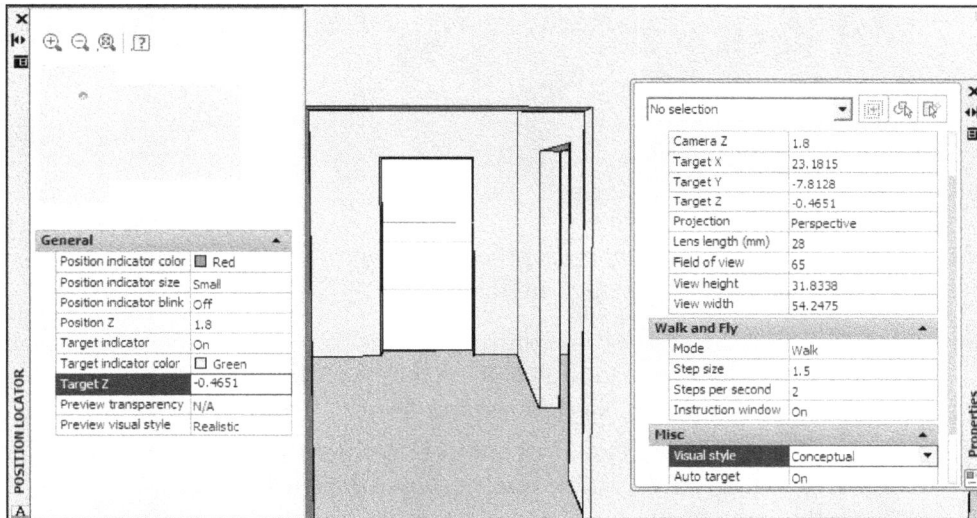

9. Press the *Esc* key to cancel the command.

10. End the drawing without saving.

Viewports

When starting AutoCAD, the drawing area is composed of a single viewport where we create, view, and modify the drawing or the 3D scene. As the project advances, it may be useful to divide the drawing area in more viewports so that we can simultaneously view different parts of the model (for instance, a bridge drawing, where we want to see both extremes) or view it from different angles. We can start a command in one viewport and end it in another. As against the layout viewports, viewports in model space never overlap and completely fill the drawing area. Any model modification will be reflected in all viewports but visualization tools are independent. We can have viewports with different zooms, pans, orbits, or orthographic views, with different visual styles, perspective, or parallel modes.

Viewport controls

The viewport controls, available in AutoCAD since Version 2012, are the three top-left menus for each viewport.

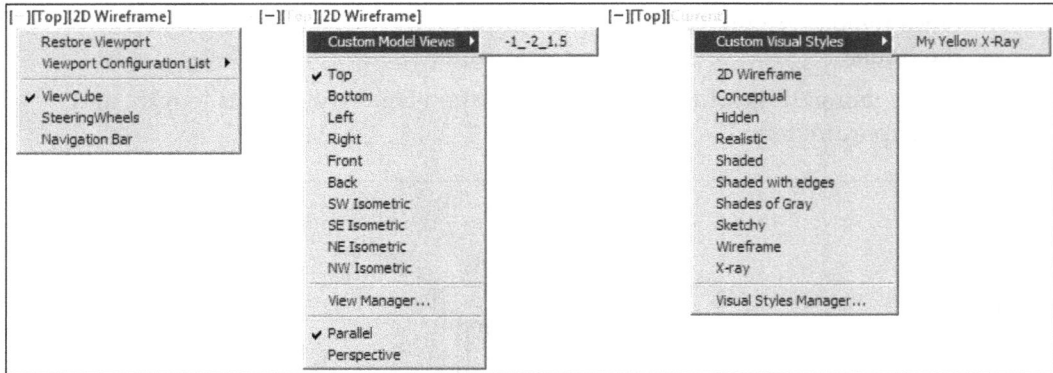

][Top][2D Wireframe]	[−][ ][2D Wireframe]	[−][Top][ ]
Restore Viewport	Custom Model Views ▸ -1_-2_1.5	Custom Visual Styles ▸ My Yellow X-Ray
Viewport Configuration List ▸	✔ Top	2D Wireframe
✔ ViewCube	Bottom	Conceptual
SteeringWheels	Left	Hidden
Navigation Bar	Right	Realistic
	Front	Shaded
	Back	Shaded with edges
	SW Isometric	Shades of Gray
	SE Isometric	Sketchy
	NE Isometric	Wireframe
	NW Isometric	X-ray
	View Manager...	Visual Styles Manager...
	✔ Parallel	
	Perspective	

These viewport controls represent easy processes for controlling multiple visualization parameters already seen in this chapter:

- **The [-] control**: This viewport control controls the viewports' configuration and activation on visualization tools such as the **ViewCube** tool and the less used **SteeringWheels** or **Navigation Bar**.

- **The [Top] control**: This viewport control specifies the view direction between the saved, orthographic, and isometric views. It is also used for applying the VIEW command and toggling between parallel and perspective modes.

- **The [2D Wireframe] control**: This viewport control specifies the visual style, and is also used for accessing the VISUALSTYLES command.

Working with multiple viewports

The VPORTS command (no alias) allows you to configure viewports. It displays a dialogue box with two tabs:

- **New Viewports**: The **New name** field allows you to name the configuration; **Standard viewports** includes all possible configurations, and the preview is displayed on the right-hand side; **Apply to** specifies if the chosen configuration is applied to the drawing area or to the current viewport; **Setup** specifies if the configuration is 2D or 3D, the last applying immediately different views. Selecting each viewport in the **Preview** area, it's possible to change the view and visual style.

- **Named Viewports**: This tab just lists and previews the named viewport configuration.

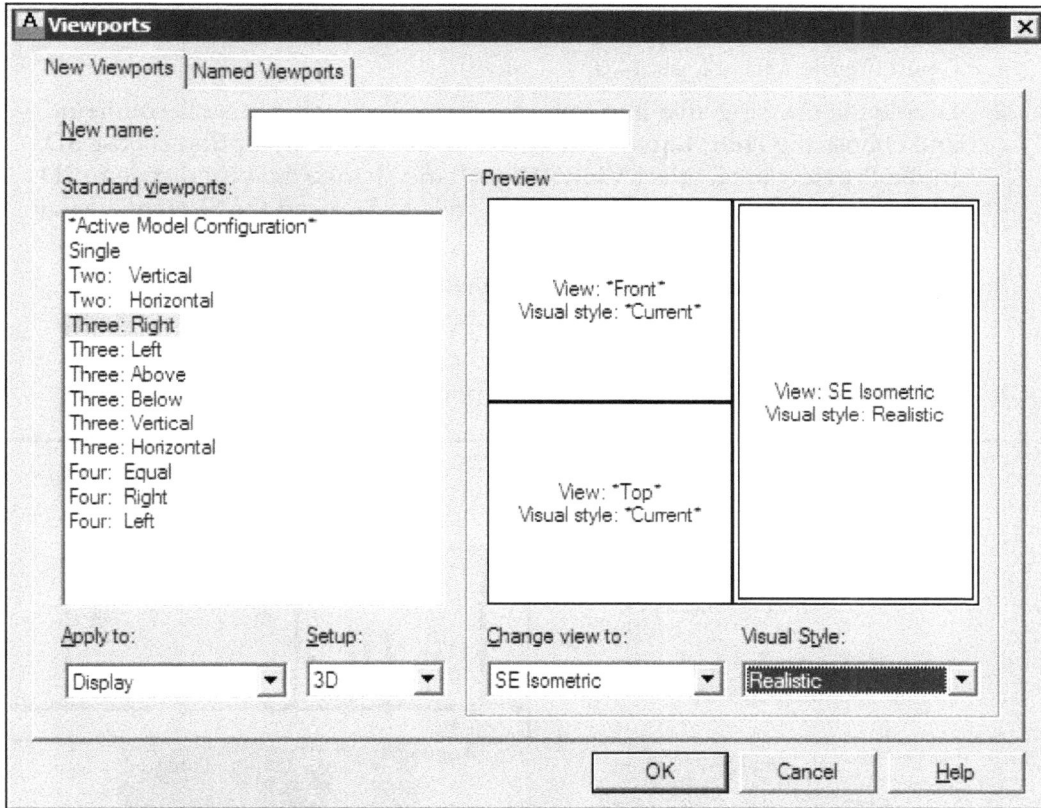

Single configuration restores a single viewport in the drawing area.

Viewport configurations can also be accessed on viewport controls, on the **Viewports** toolbar, the **Viewports** menu bar in the **View** tab, and on the ribbon's **View | Model Viewports** panel.

> As with most commands that display dialogue boxes, VPORTS also has a command-line version, -VPORTS. This one has the **Toggle** option, which allows for a quick change between a single viewport and multiple viewports.

Exercise 2.5

We are going to divide the drawing area, set different views, and save this viewport's configuration.

1. Open the file A3D_02_05.DWG.

2. Divide the drawing area into four viewports. Apply the VPORTS command and choose the **Four: Equal** configuration. On the **Setup** option choose **3D**. In the **Preview** area, select **View:*Right*** and change the visual style to **2D Wireframe**. Apply the **Realistic** visual style to **Top** and the **Sketchy** visual style to **Front**.

3. Still within the dialogue box, select **View: SE Isometric** and change the view to ***Current***. Name this configuration **My Viewports** and click on the **OK** button.

4. Adjust the zoom and pan in each viewport.

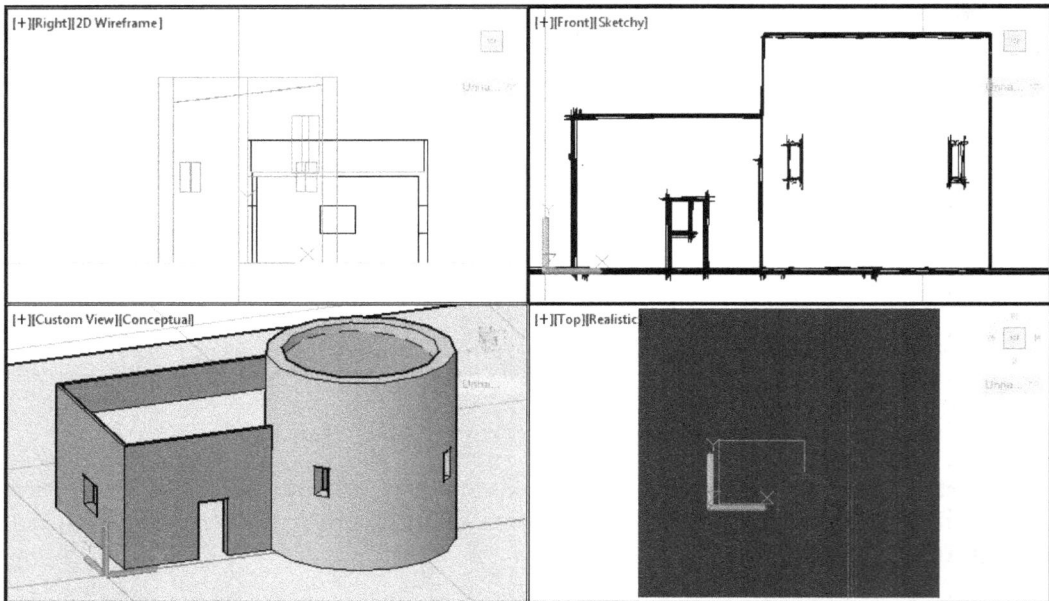

5. Apply the -VPORTS command and select the **Toggle** option. We come back to a single viewport.

6. Quit the drawing without saving.

Summary

In this chapter, we analyzed all important aspects and commands for a correct visualization of 3D models. We started with zoom, pan, and orbiting, then we saw predefined and saved views and the creation of cameras and their parameters. We explored the perspective mode in order to get a more realistic visualization, and the visual styles that allow for viewing the model in the wireframe, shaded, shaded with materials (realistic), and other modes. We learned how to walk and fly through the model and we saw how to divide the drawing area in several viewports.

In the next chapter we will learn how to create coordinate systems, thus defining different XY working planes.

3
Coordinate Systems

Along with visualization, the creation of other working planes is imperative for proper 3D modeling. Frequently, we have to create objects that are not aligned with the current working plane. Instead of creating and aligning them to the final position, sometimes it is much faster to define the correct working plane and only then proceed to create objects. We must not forget that an important part of AutoCAD commands work only in 2D (the XY plane).

In this chapter, we will cover:

- The world coordinate system
- How to create and apply user coordinate systems
- The UCS command
- Creating UCS by moving and orienting the icon
- When and how to apply dynamic coordinate systems
- Coordinate system icons
- Important variables to know

User coordinate systems

We present the world coordinate system as the AutoCAD reference coordinate system, and we shall discuss how to create user coordinate systems.

The world coordinate system

The world coordinate system, also known as World or WCS, is the reference coordinate system and it can never be deleted or modified. All user coordinate systems are based on this one. When we create an object, it will be internally defined to the WCS, even if it is not the current coordinate system.

The WCS is also the coordinate system used when exporting the 3D project to other formats.

[When we start a 3D project, it is wise to start it near the WCS origin, with its height aligned to the WCS Z axis.]

Creating user coordinate systems

Following are the important commands and variables related to user coordinate systems, also known as UCS. The main command is aptly named ucs.

The UCS command

The ucs command (no alias, but it's easy) allows us to create and manage user coordinate systems. There are several options for creating a UCS: from the UCS toolbar, by going to **Tools | New UCS** on the menu bar, the **Coordinates** panel in the **Home** tab on the ribbon, or right-clicking over the UCS icon.

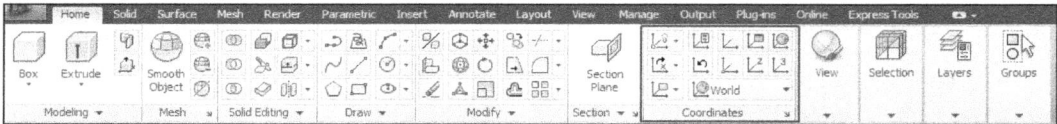

The command that can be seen on the command line informs us about the current UCS and, by default, creates a UCS defined by three points, starting with the new origin:

```
Command: UCS
Current ucs name: *WORLD*
Specify origin of UCS or [Face/Named/Object/Previous/View/World/X/Y/Z/
ZAxis] <World>: Point
```

Then, it asks for a point that defines the new X positive direction. If we press the *Enter* key, the command ends and we have created a UCS that has a different origin but maintains the same directions for the X and Y axes:

```
Specify point on X-axis or <accept>: Point
```

With these two points, which define the origin and the X axis, an infinite number of planes are possible. So we need to specify a third point to define the XY working plane. If we press the *Enter* key, the command ends and we have got a working plane parallel to the previous one.

```
Specify point on the XY plane or <accept>: Point
```

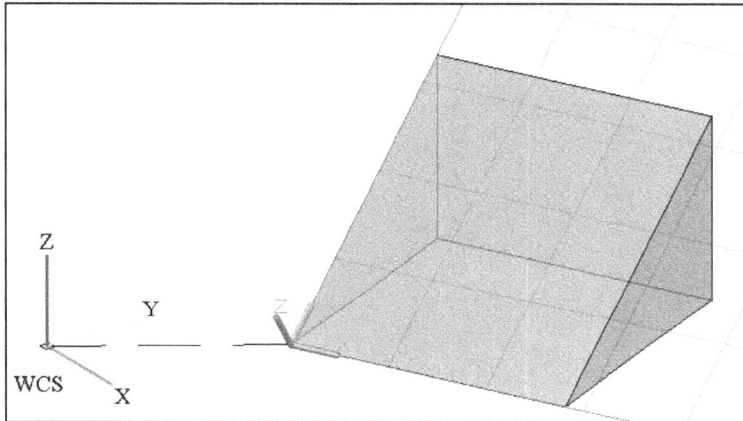

We must be very careful when clicking on these points and be sure that the correct object snap is assigned. A wrong UCS, if not immediately detected, can cause serious problems.

Besides defining a UCS by three points, other creating options are available, as follows:

* **Face**: It creates a UCS aligned with any planar face of a solid surface or mesh. We click inside the face that will define the UCS. The origin is the clicked point and the X direction is parallel to the closest edge. Optionally, we can select the next face according to the point projection by flipping around the X axis or the Y axis:

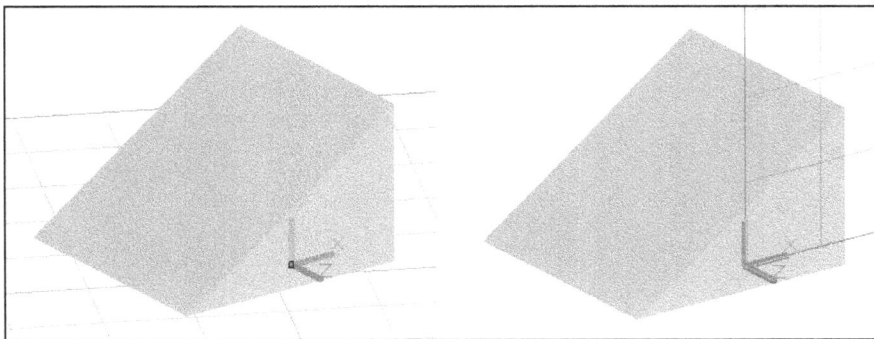

- **Object**: It creates a UCS defined by an existing object. For instance, selecting a line creates a UCS parallel to the one used to create the line, with the origin at the endpoint closest the selection point and the X axis given by the line.

- **View**: It creates a UCS perpendicular to the viewing direction, meaning that the X and Y axes are aligned with the drawing area. The origin doesn't change.

- **X / Y / Z**: It creates a UCS by rotating the current UCS around the X, Y, or Z axis with the specified angle value. The origin doesn't change:

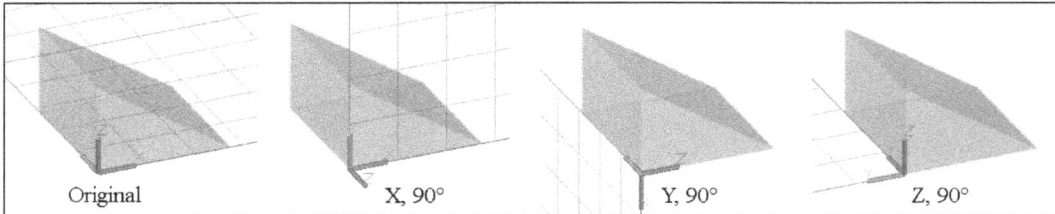

| Original | X, 90° | Y, 90° | Z, 90° |

- **Z Axis**: It creates a UCS defined by an origin and its Z direction and is useful when we want a working plane perpendicular to a particular direction.

The **Previous** option activates the UCS that was last used. This option can be used successively to recover the last 10 UCSs.

For later use, it is also possible to save and recover the UCS. Applying the **Named** option, we get a new set of options: **Save** to save this UCS (a name is requested), **Restore** to activate a saved UCS (its name is requested), **Delete** to eliminate the definition of a saved UCS, and **?** (question mark) to list all saved UCSs.

To activate the world coordinate system, upon entering the UCS command we just press the *Enter* key or click on the **World** option.

> This command has two hidden options. The **Apply** option applies the current UCS in the active viewport to all other viewports. The **orthoGraphic** option allows for activation of an orthographic UCS, related to the world coordinate system and obtained by rotating this one. It requests the UCS with the following:
>
> ```
> Enter an option [Top/Bottom/Front/Back/Left/
> Right] <Top>:
> ```

Creating a UCS by moving and orienting the icon

Since Version 2012, it has been possible to create a UCS by moving and orienting its icon directly. We select the icon and four grips are displayed (if in a 3D View), one at the origin and the other three at each axis's endpoint. Any of these grips can be activated and moved, thus allowing for a dynamic creation of coordinate systems:

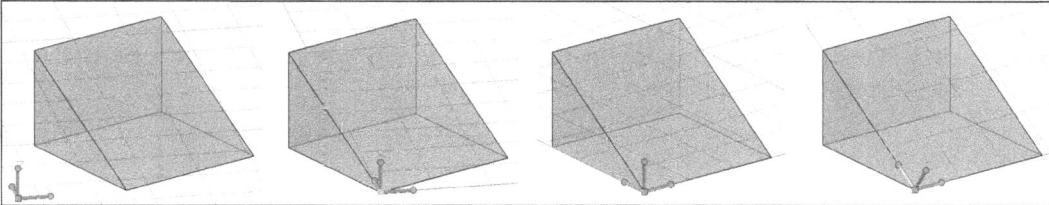

On right-clicking over the UCS icon, all creation options as well as a list of saved UCSs are displayed.

Important variables to know

AutoCAD has several variables related to the coordinate systems, which affect its behavior when creating a UCS or changing a view direction. Here are the most important ones:

- UCSFOLLOW: On activating a UCS while keeping the value of UCSFOLLOW as 1, the view changes to the respective plan view. It has a value of 0 by default and is saved in the drawing.

- UCSVP: When working with more than one viewport, this variable allows for having an independent UCS per viewport (value 1) or only one UCS in all viewports (value 0). It has a value of 1 by default and is saved in the drawing.

- UCSORTHO: It controls whether the corresponding orthographic UCS is activated (value 1) or not (value 0) when changing the view to an orthographic view. It has a value of 1 by default and is saved in the registry.

- UCSDETECT: It controls if dynamic UCS is on (value 1) or not (value 0). It has a value of 1 by default and is saved in the registry.

- UCSSELECTMODE: It controls whether the UCS icon can be selected, moved, and oriented (value 1), or not (value 0). It has a value of 1 by default and is saved in the registry.

Exercise 3.1

We will draw some linear objects in different planes of a 3D model included in the accompanying files.

1. Open the A3D_03_01.DWG file. It represents a part with several planes that are not aligned with the WCS.

2. Create a UCS aligned with the front face. Apply the UCS command and click on the lower-left corner, lower-right corner, and any upper vertex on that face, in the said sequence:

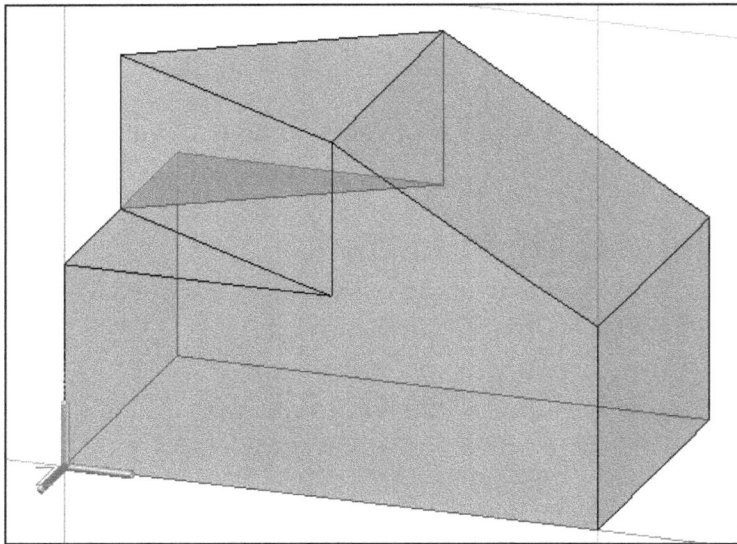

3. Apply the LINE command, pause on the lower-left corner (actually the origin of the new UCS), aim at the X axis, and type in 2 to specify the first point distanced 2 units from the corner. Then aim at the Y axis (up) and type in 2, at the X axis and type in 1, and at the Y axis (down) and type in 2. End the command.

4. Apply the RECTANG command (alias REC), with the first corner in absolute coordinates. Type 4,1 as the first corner and @1.4,1 as the opposite corner:

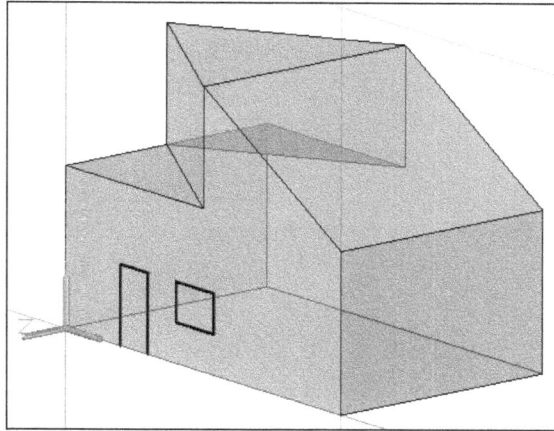

5. Create a UCS aligned with the right face, again using three points.

6. Apply the RECTANG command to create a rectangle from 1,1 (absolute coordinates) to @4,1 (relative coordinates).

7. Now, activate the UCS icon and position and orient it with the top inclined face by moving its grips. Users with versions older than 2012 should apply UCS using three points again.

8. Create a rectangle coincident with the face edges, either with RECTANG or PLINE.

9. Apply the OFFSET command (alias O) to the last object (easily selectable by typing L on the command line) and specify a distance of 0.2 and mark a point inside the object:

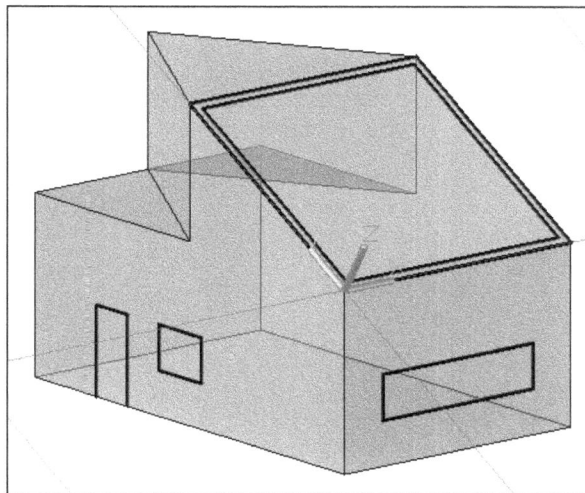

10. Create a UCS by rotating the current one 90 degrees around the X axis. Use the UCS command and select the **X** option. Accept 90 as the default value.

11. Apply the PLINE command (alias PL). The first point is the origin. With **Ortho** on, mark 2 along Y, 6 along X, and -2 along Y (or specify the face endpoint):

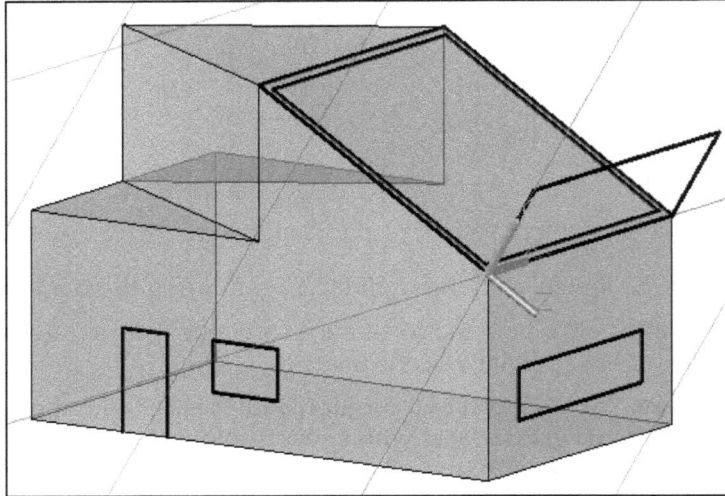

12. Save the current UCS with the name Top perpendicular.

13. Create a UCS on the upper-front inclined face, using either the icon or three points.

14. Create a rectangle coincident with that face edges.

15. Apply the OFFSET command (alias O) to the last object, specify a distance of 0.2, and mark a point inside.

16. Do the same to the upper-posterior inclined face.

17. Activate the world coordinate system by, for instance, applying UCS and pressing the *Enter* key:

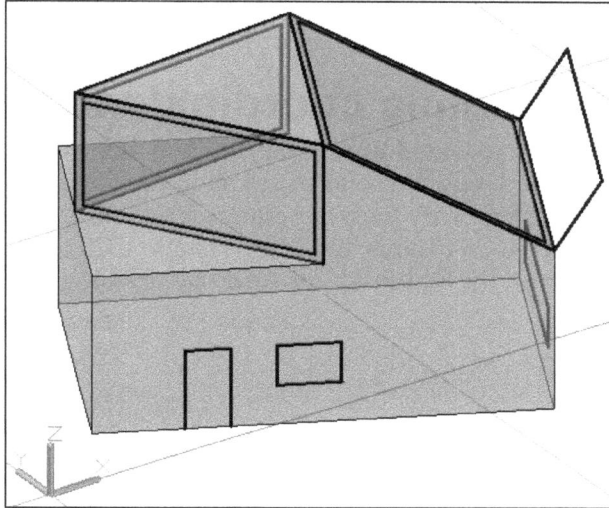

18. Save the model with the name A3D_03_01final.DWG.

Managing user coordinate systems

There are some ways to manage saved user coordinate systems. Besides the UCSMAN command, two other ways are the small list below the view cube and the UCS icon context menu under **Named UCS**.

The UCSMAN command

The UCSMAN command (alias UC) allows you to manage saved UCSs. It displays a dialog box with three tabs:

- **Named UCSs**: We can activate **World**, **Previous**, or a saved UCS.
- **Orthographic UCSs**: We can activate one of the six orthographic UCSs.
- **Settings**: It contains options related to the UCS icon, indicating if it is displayed, if it is displayed at the origin, if these settings are applied to all viewports, and if it is selectable. Also, it contains two options related to the UCS: if it is saved with the viewport and if the Plan view is activated when changing the UCS.

Dynamic coordinate systems

Since Version 2007, an auxiliary tool is available for creating objects on a solid planar face.

Creation of dynamic coordinate systems

If the dynamic UCS tool is activated during the creation of 2D or 3D objects, then when moving the cursor over a solid planar face, the face is highlighted and AutoCAD creates a temporary working plane on that face. When ending the command, the temporary UCS disappears. The dynamic UCS can be turned on or off by the **DUCS** button on the Status bar, or easily by pressing the *F6* key:

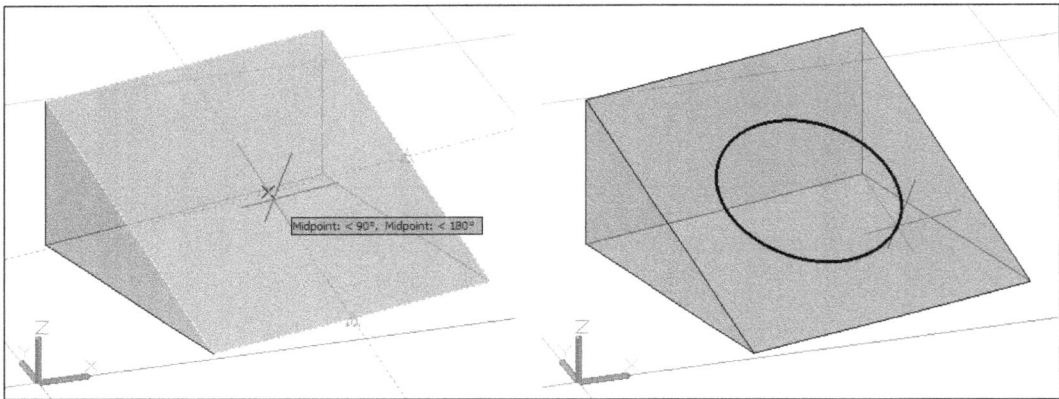

> Normally, the object's position on that plane is not at a precise distance from an edge or vertex. One way to have precision is to turn on **OTRACK**, pause over one or two object snaps, and direct the cursor. Another handy auxiliary tool is FROM, a special object snap available on the individual osnap menu (*Ctrl* and right-click). This tool allows you to specify a point with relative coordinates from a base point.

Exercise 3.2

With a dynamic UCS, create some more objects in the 3D model from the last exercise.

1. Open the A3D_03_02.DWG file.

2. Turn on **DUCS**.

3. Create some circles and rectangles on the front and left faces:

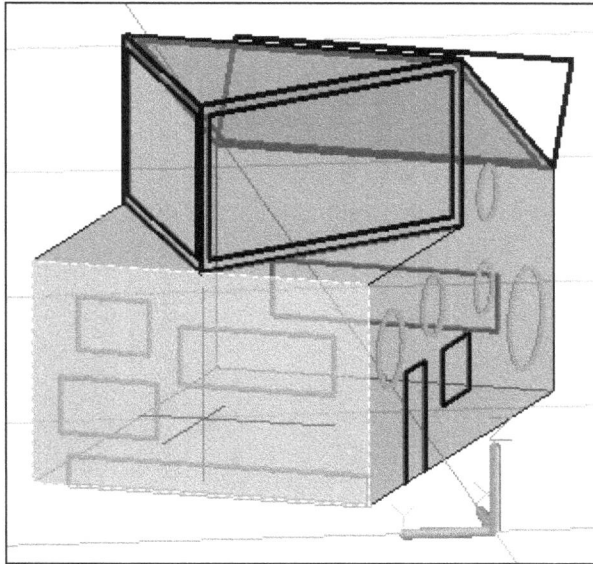

4. Create a square of side 1 on the upper inclined face from the lower-left corner of the inner rectangle, 2 units along the horizontal and 1 unit along the vertical. Apply the RECTANG command, pause over the face, press *Ctrl*, and right-click and select **From**. Specify the lower-left corner of the inner rectangle as the base point and type @2,1 for the offset distance. Then type @1,1 as the opposite corner:

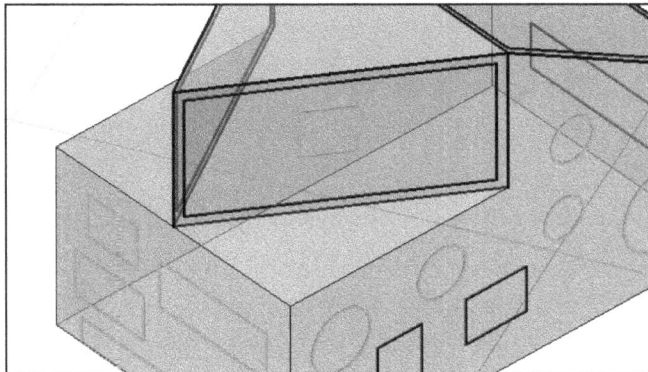

5. There is no need to save the model.

UCS icons

In 3D, the UCS icon, represented by the three orthographic axes, is very important as it tells us every instant which and where is the active UCS. Normally we work with a 3D icon, but a 2D icon is also available.

Identifying UCS icons

Depending on the situation and current visual style, the UCS icon can adopt different aspects:

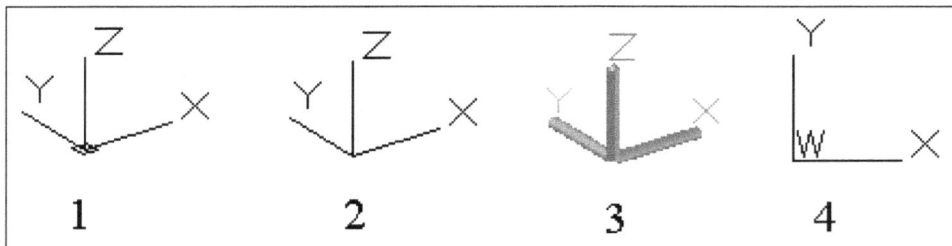

In 2D wireframe visual style, the icon is represented by lines. If the UCS icon has a small square (as shown in image 1), it means that the world coordinate system is current. Without the small square (as shown in image 2), it means it's a current user coordinate system. With all visual styles, except 2D wireframe, the icon is shaded (as shown in image 3) and there is no difference between the world coordinate system and the user coordinate system. Particularly in 2D projects, we may opt for a 2D icon (as shown in image 4) where the Z axis is absent and the world coordinate system is indicated by the letter W.

Modifying the UCS icon

To display or modify the UCS icon, we can apply the UCSICON command. It has the following options:

- **ON/OFF**: This turns the UCS icon visibility on or off in the current viewport.
- **All**: This will apply visibility and origin settings to all viewports.
- **Noorigin/ORigin**: This displays the icon at the coordinate system origin if the origin is visible in the current viewport. If the origin is not visible, it is displayed at the lower-left corner.
- **Selectable**: This specifies whether the UCS icon is selectable or not (this option is available since the 2012 version).

- **Properties**: This displays a dialog box to control the icon aspect. We can choose between a 2D or 3D icon and the line width, the icon size, and the icon colors in the model space and layouts. With **Apply single color** checked, the icon has a uniform color. Otherwise it applies the same colors as shaded in the previous figure: red for the X axis, green for the Y axis, and blue for the Z axis:

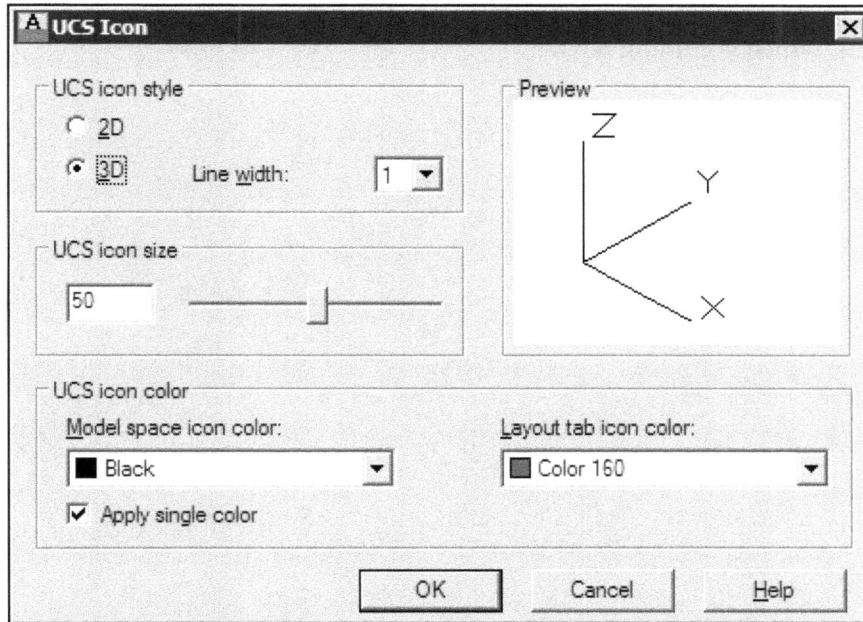

Summary

In this chapter we learned multiple methods for creating user coordinate systems, including the ucs command, and also covered manipulating the UCS icon and activating the dynamic automatic UCS. User coordinate systems are fundamental for accessing different working XY planes. We now know how to manage the saved UCS and to identify and change the UCS icon. We also view the most relevant variables that control UCS behavior. Now, we have the basis for the creation of real 3D models, which is coming next.

4

Creating Solids and Surfaces from 2D

Finally, we've come to the main commands for creating 3D models. The commands explained in this chapter allow you to create solids or surfaces from linear or planar objects.

In this chapter we will cover:

- How to create solids and surfaces by extrusion
- How to quickly create solids from closed areas
- How to create solids and surfaces by revolution
- How to create solids and surfaces with different sections
- How to create solids and surfaces along a path
- Important variables related to solids and surfaces

Creating solids or surfaces

We can create solids from 2D closed objects, such as circles, ellipses, closed polylines (including rectangles and polygons), and closed splines. Additionally, we can create solids from regions, 3D faces, 2D solids (created with the SOLID command), planar faces of 3D solids, planar surfaces, and traces (an obsolete entity created with the TRACE command).

We can create surfaces (3D objects without thickness) from open linear objects, such as arcs, lines, elliptical arcs, open polylines, and open splines. We can also create surfaces from edges of 3D solids and surfaces.

All the commands explained in this chapter, except PRESSPULL, have the **MOde** option. This option allows you to choose between **Surface** and **Solid** and is useful when we want to create surfaces from closed objects. Also, the **Expression** option, allows the introduction of a formula or mathematical expression whenever a value is requested, and is transversal to many AutoCAD commands.

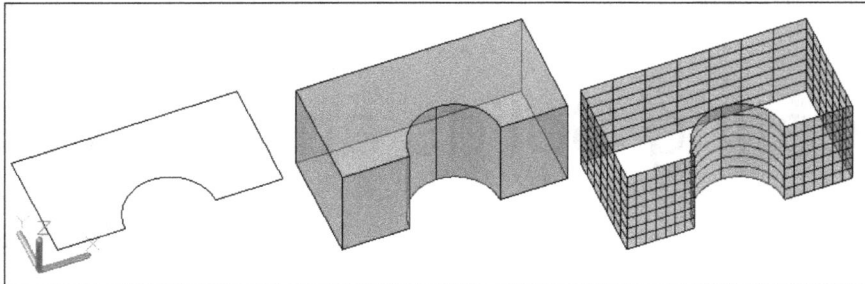

All commands are available on the ribbon, such as the **Home** tab or the **Modeling** panel, and the **Modeling** toolbar, the **Draw/Modeling** menu bar (with the exception of **PRESSPULL**).

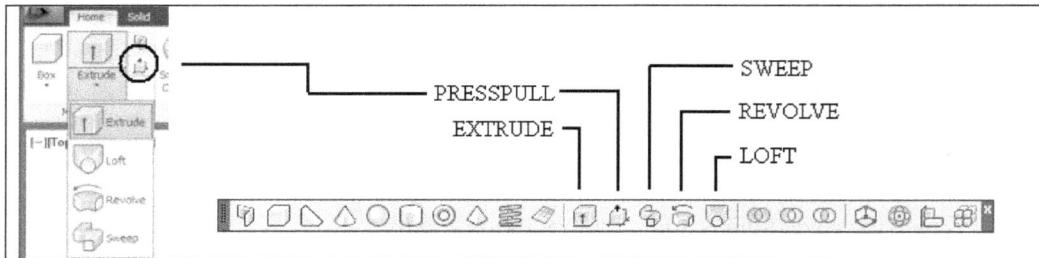

Creating solids and surfaces by extrusion

Often we create a 3D model from 2D drawings by extruding (giving height) to planar entities or closed areas. For instance, in architecture, we may construct walls from plan views. The following sections present the commands for extrusion.

Extruding 2D objects

Probably the most used command for modeling in 3D is EXTRUDE, which is presented next.

The EXTRUDE command

The EXTRUDE command (alias EXT) allows for extruding or giving height to planar entities, thus creating volumes. The command starts by showing information about the wire frame density and the **MOde** default option; it then asks for the objects to be extruded:

```
Command: EXTRUDE
Current wire frame density:  ISOLINES=4, Closed profiles creation mode
= Solid
Select objects to extrude or [MOde]: Selection
```

By default, the command prompts the height of extrusion:

```
Specify height of extrusion or [Direction/Path/Taper angle/
Expression]: Value
```

All the selected objects are extruded. If the original object is closed, a solid is created, else a surface is created. By default, when creating a solid, the original object is erased. This behavior is controlled by the DELOBJ variable.

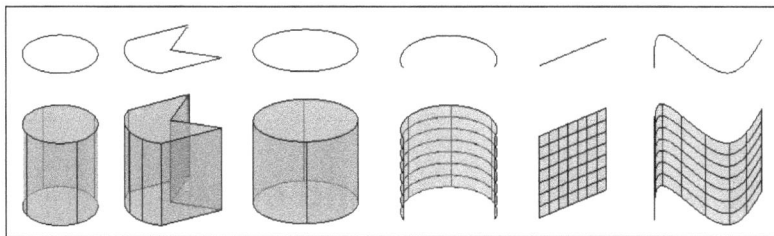

If the object is closed but we want to create a surface, we must change the **MOde** option to **Surface** before proceeding to the selection.

Other command options are as follows:

- **Direction**: This option creates an inclined extrusion. It prompts for two points that specify length and angle:

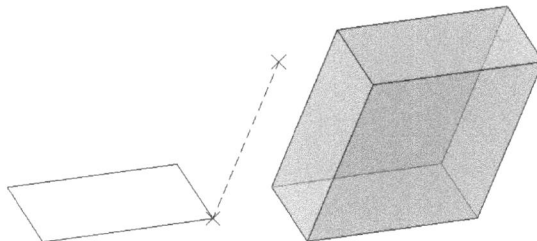

- **Path**: This option creates an extrusion along a path. It prompts for the selection of a linear object.

> Instead of using this option, I advise you to apply the SWEEP command, which provides greater flexibility and better results for complex geometry.

- **Taper angle**: This option allows you to create an extrusion object with an angle applied to the vertical faces. It prompts you for the taper angle and height. If the angle is positive, sections decrease; and if negative, sections increase. If the taper angle or height values are too large, a vertex or edge may be reached before the specified height.

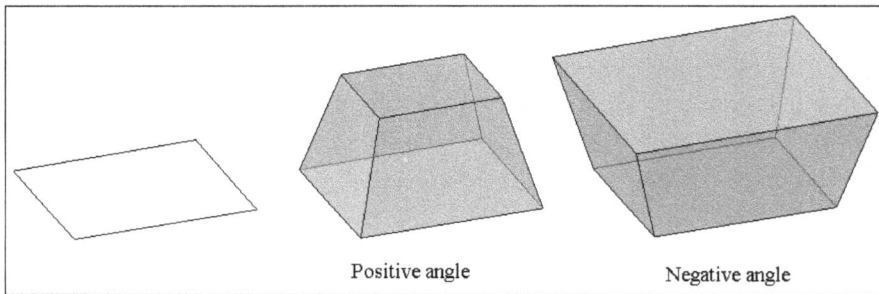

Positive angle   Negative angle

EXTRUDE applied to faces and edges

If we press the *Ctrl* key, AutoCAD allows us to select the vertices, edges, and faces (called subobjects) of solids and surfaces.

So, we can apply EXTRUDE to edges and faces. When prompting for the selection, we press the *Ctrl* key and select these subobjects. Then, we continue the command by typing the height value or apply another option. The new objects are independent of the old ones.

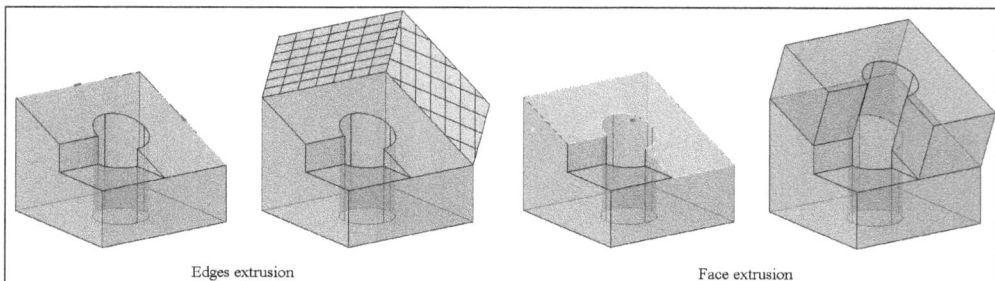

Edges extrusion   Face extrusion

Exercise 4.1

Do you remember the house from the *Exercise 2.1* section of *Chapter 2, Introduction to 3D Design*? It is now time to start its creation from the 2D plan by following these steps:

1. Open the file A3D_04_01.DWG.

2. Let's start by creating the 3D layers. Using the LAYER command (alias LA), create the layers (colors at your discretion), such as 3D-WALLS, 3D-SLABS, 3D-BALCONY, and 3D-OPENINGS. Activate the first one and freeze the WINDOWS layer.

3. Applying the BOUNDARY command (alias BO), choose **Region** as **Object type** and click on **Pick Points**. Then click inside all walls and press the *Enter* key. The five regions should be created.

4. With the third viewport control, or the VSCURRENT command (alias VS), change the visualization style to **Realistic**. All walls should be opaque; if not, polylines will be created instead of regions:

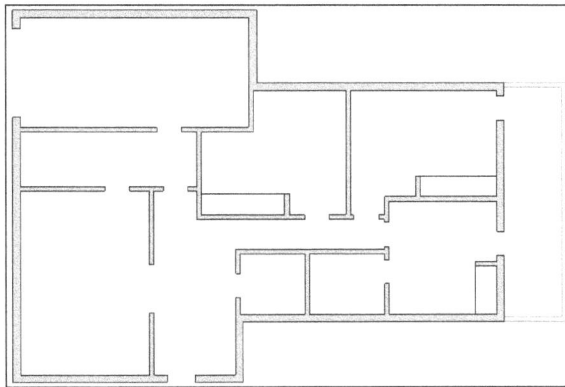

5. We are going to extrude these regions, but first, on the layers list, freeze WALLS. In this way, we will avoid extruding the wrong objects. Apply the EXTRUDE command (alias EXT), select all five regions, and specify a height of 2.7 units. When orbiting the view, you can already see 3D walls.

6. Activate the 3D-BALCONY layer, and, again using the BOUNDARY command, create a region inside the balcony lines.

7. Then freeze the BALCONY layer and extrude the balcony region by 1 unit.

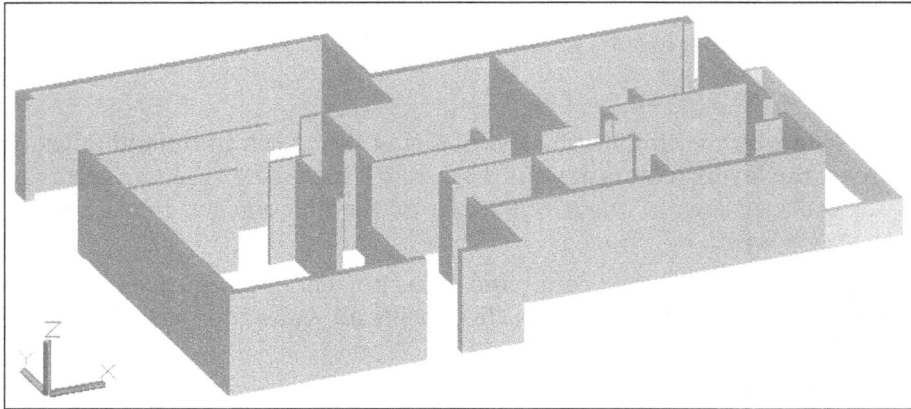

8. To create the parts of the doors shown in the previous screenshot, activate the 3D-WALLS layer. To each door opening, create a rectangle or close polyline using object snaps.

9. Select all the rectangles or polylines (it is easy if selected on the front view of a window) and apply EXTRUDE with height -0.7. By specifying a negative value, extrusions are processed in the negative Z direction. In *Chapter 7, Editing Solids and Surfaces* we will unite these elements to walls.

10. To create window openings, activate the 3D-OPENINGS layer, freeze 3D-WALLS and 3D-BALCONY layers, and thaw the WINDOWS layer.

11. We will once again create rectangles or closed polylines contouring each 2D window.

12. Before extruding, freeze the WINDOWS layer. Then apply EXTRUDE to all visible rectangles or closed polylines, with height as 1 unit.

13. Move all future openings by 1 unit in the Z direction, thus positioning these in the proper place. In *Chapter 7, Editing Solids and Surfaces*, we will make the subtractions.

14. To create the first slab, activate the 3D_SLABS layer and then thaw the WALLS and BALCONY layers.

15. Create a closed polyline contouring all 2D geometry.

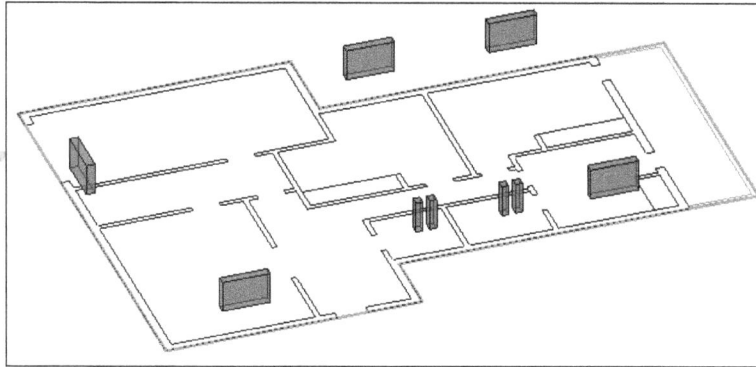

16. Apply EXTRUDE to the last object created with height -0.1.

17. Let's unfreeze the 3D-WALLS and 3D-BALCONY layers and freeze WALLS and BALCONY.

18. To create the top slab, we could use the same process of contouring the top of the walls, but another easy way is to copy the existing slab and apply the SLICE command to cut the part corresponding to the balcony. We will apply this last method. For now, select the slab and copy it 2.8 units along Z direction. In *Chapter 7, Editing Solids and Surfaces*, we will conclude this part.

19. By changing the visualization style to **X-Ray** and orbiting, verify the model's accuracy.

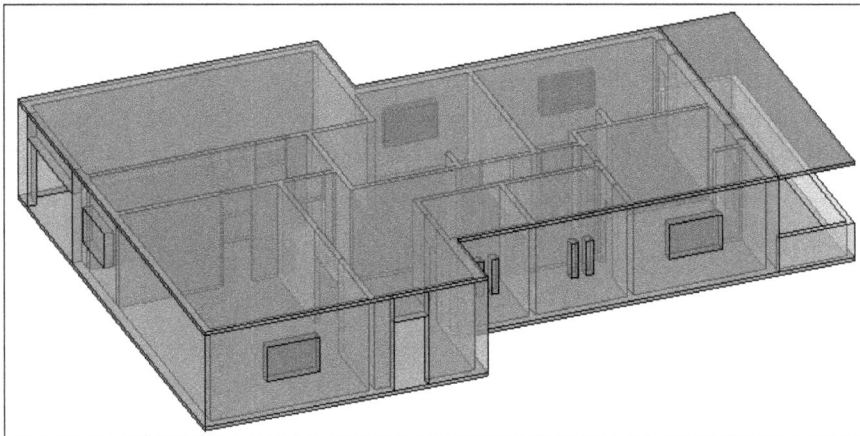

20. Save the model with the name A3D_04_01FINAL.DWG.

Creating solids from closed areas or faces

Presented next is another suitable command to create solids from closed areas or faces.

The PRESSPULL command

PRESSPULL is a command that allows us to create solids dynamically and in a simpler way. Since Version 2013, it also allows us to create surfaces:

- **Solids**: This creates a solid by selecting a closed object or specifying an internal point to a closed area and then entering a value or picking a point to specify the height
- **Surfaces**: This creates a surface by selecting an open 2D object

Since Version 2012, the command repeats automatically, allowing for multiple extrusions. Pressing the *Enter* or *Esc* key ends the command.

The most common situation is when we have several closed areas and we want to their define volumeter quickly.

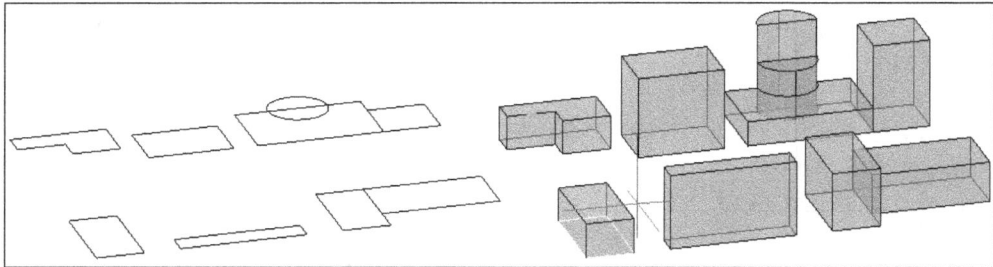

Version 2013 introduced one additional possibility. By pressing the *Shift* key, we can specify several closed areas. Then press the *Enter* key and indicate a uniform height.

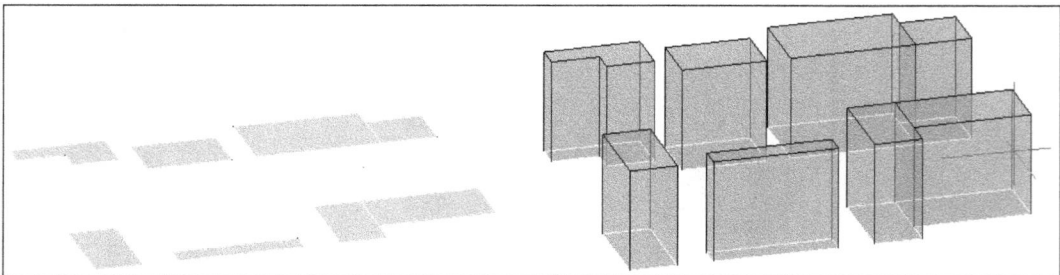

PRESSPULL applied to faces

This command also allows us to extrude the planar faces of existing solids. Upon entering the command, we can select faces of solids. Then, the command accepts either a point or a value for the height of the extrusion. And, as with areas, if we press the *Shift* key, we can select several faces and apply the same height.

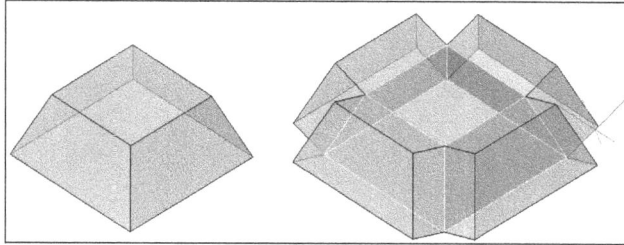

Another innovation in Version 2013 is the difference between extruding and offsetting faces. If the face is normally selected, it is extruded while maintaining the area, but if the face is selected while keeping the *Ctrl* key pressed, the face is offset and the adjacent face maintains its angle.

Exercise 4.2

From a 2D plan, we will create the volumetric representation of a street in a couple of minutes by following these steps:

1. Open the file A3D_04_02.DWG. This simple drawing has several closed areas, each with a number that represents the height of extrusion.

2. Before applying the command, orbit the drawing:

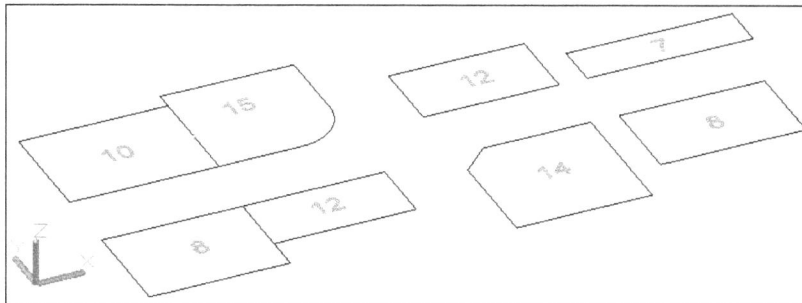

3. Apply PRESSPULL, click inside the first area, and type its number as height. The command repeats and we continue clicking inside each area and typing the respective height. When finished with all areas, end the command.

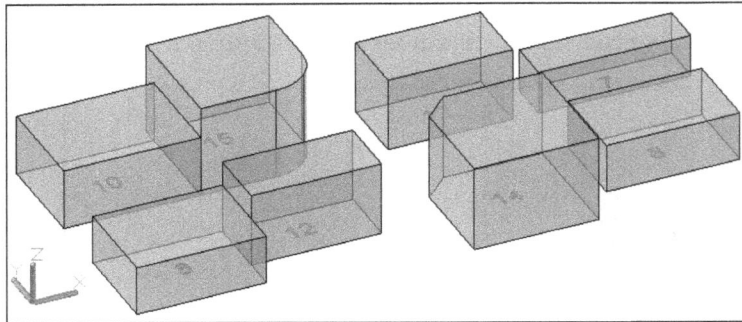

4. There is no need to save the drawing.

Creating solids and surfaces by revolution

When we need a solid or surface whose section is rotated around an axis, the next command should be applied.

The REVOLVE command

The REVOLVE command (alias REV) creates a solid or surface by rotating a linear or planar object around an axis. It starts by showing the same information and prompts for the selection of sections, as seen with EXTRUDE. We can select any planar object. If an object is not closed or if the **MOde** option is set to **Surface**, a surface is created:

```
Command:  REVOLVE
Current wire frame density:  ISOLINES=4, Closed profiles creation mode
= Solid
Select objects to revolve or [MOde]: Selection
```

We must then define the axis of revolution. By default, we can pick two points that define this axis, but other options are available as well. In the order we select, the points specify the positive angle direction by the right-hand rule.

```
Specify axis start point or define axis by [Object/X/Y/Z] <Object>:
Point
Specify axis endpoint: Point
```

Finally, type the angle of revolution, which is the default value. By typing a negative value, the angle is measured in the opposite direction.

```
Specify angle of revolution or [STart angle/Reverse/EXpression] <360>:
Value
```

The result of the previous code snippet is shown in the following screenshot:

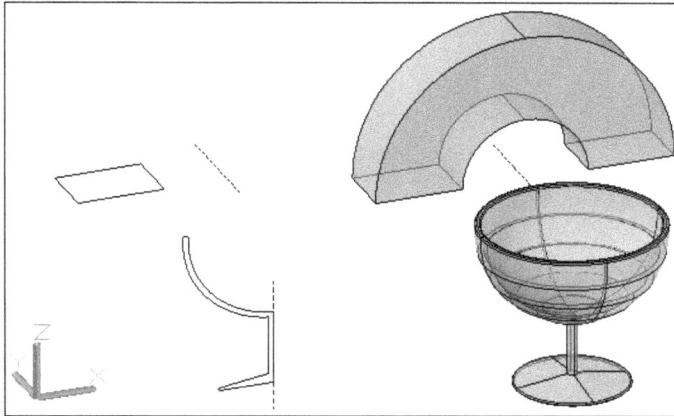

The other command options, either for axis or revolution angle definition, are:

- **Object**: We can select an object to define the axis of revolution. Eligible objects are lines, segments of polylines, and edges of solids or surfaces (if the *Ctrl* key is pressed).

- **X/Y/Z**: The axis of revolution is defined by the positive direction of the X, Y, or Z axis of the current coordinate system.

- **STart angle**: The revolution object starts at this angle from the section object.

- **Reverse**: It allows us to reverse the direction angle. It is similar to specifying a negative angle value.

> An interesting curiosity is that when the command prompts for the angle of revolution and if we press the *S* key, it applies an angle of 270 degrees.

Exercise 4.3

We are going to create a circular balcony by following these steps:

1. Open the file A3D_04_03.DWG. This drawing already includes all needed sections.

2. Apply the EXTRUDE command to the semicircle, with height as -0.15:

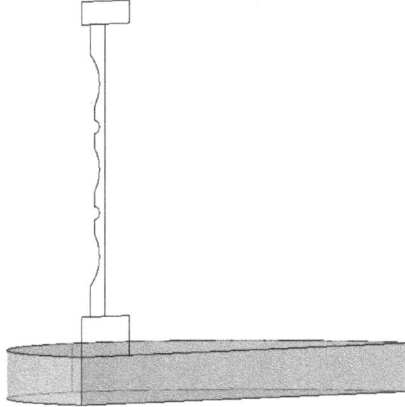

3. To create the hand rail and base rail, apply the REVOLVE command to both rectangles. The first point of the axis is the midpoint or center of the semicircle, and the other point of the axis is any point along the positive Z direction (**Ortho** or **Polar On**). The angle of revolution is -180 degrees.

4. The first baluster is also obtained from the application of the REVOLVE command to the vertically closed polyline. The axis is defined by both endpoints of the vertical segment and the angle is 360 degrees.

5. To create the other balusters, apply the ARRAY command, specifying **Polar**, the center at the midpoint or center of semicircle element, 11 elements, angle to fill -180 degrees, and rotating objects as they are copied.

> Since Version 2012, the ARRAY command changed considerably, so for these situations the ARRAY command is faster. With Version 2013, we could also apply the new ARRAYCLASSIC command, which is the same as the ARRAY command in versions prior to 2012.

6. The two balusters at the ends are erased. We could also apply the SLICE command (presented in *Chapter 7, Editing Solids and Surfaces*) to cut half of each one.

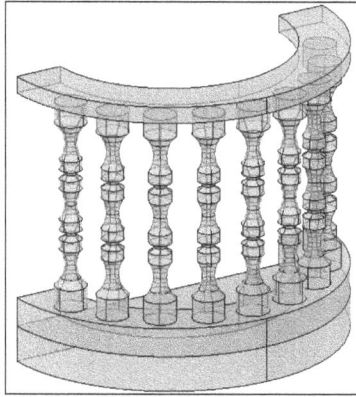

7. Save the drawing with the name A3D_04_03FINAL.DWG.

Creating solids and surfaces with different sections

Sometimes, we need to create some non-regularly shaped elements. The next command represents an easy way, and is also applied for the creation of terrains, curtains, and so on.

The LOFT command

The LOFT command (no alias) creates a solid or surface from the selection of two or more cross sections. If all sections are closed, the result is solid, unless the **MODe** option is set to **Surface**.

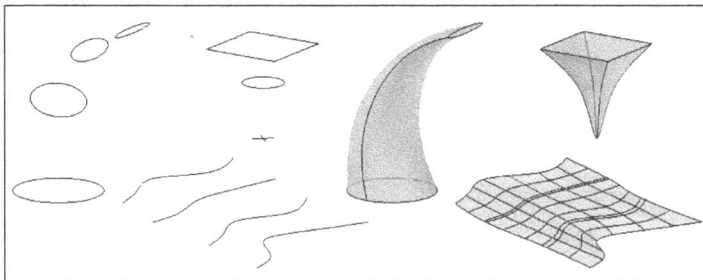

The command shows the same information as in previous commands. It then asks for the objects that define the cross sections in the correct order. As sections are being selected, the lofted object is being previewed. When ending the selection, we press the *Enter* key. The command informs us of the number of sections selected and continues:

```
Command: LOFT
Current wire frame density:  ISOLINES=4, Closed profiles creation mode
= Solid
Select cross sections in lofting order or [POint/Join multiple edges/
MOde]: Selection
n cross sections selected
```

After selecting all cross sections, the command previews the object and displays a small triangular grip that shows options for the shape of the object being created.

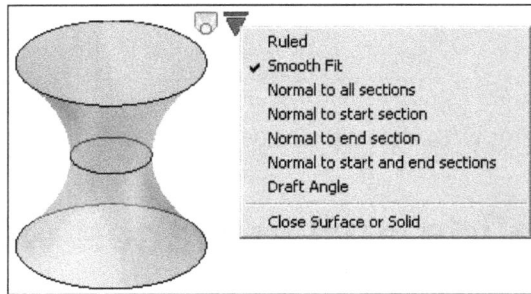

Also, some options are displayed on the command line. Pressing the *Enter* key ends the command:

```
Enter an option [Guides/Path/Cross sections only/Settings] <Cross
sections only>: Enter
```

Considering the lofted shape, we can choose between **Ruled, Smooth Fit, Normal To All Sections, Normal To Start Section, Normal To End Section, Normal To Start And End Sections**, and **Draft Angle**.

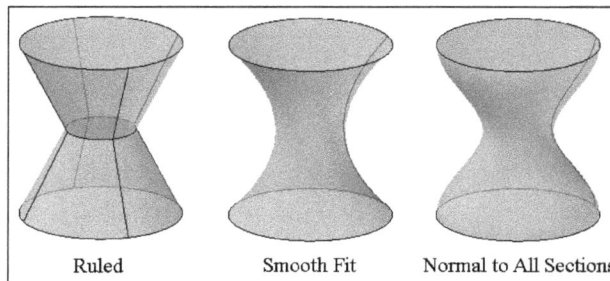

| Ruled | Smooth Fit | Normal to All Sections |

After creation, it is very simple to modify this lofted solid or surface. By modifying it without using the LOFT command, we can access the same small triangular grip, allowing us to change the lofted shape or any grip from the original sections. Another convenient way to edit the lofted object is with the **PROPERTIES** palette. This last method is the best way, with **Use draft angles**, to adjust both draft angles and magnitudes.

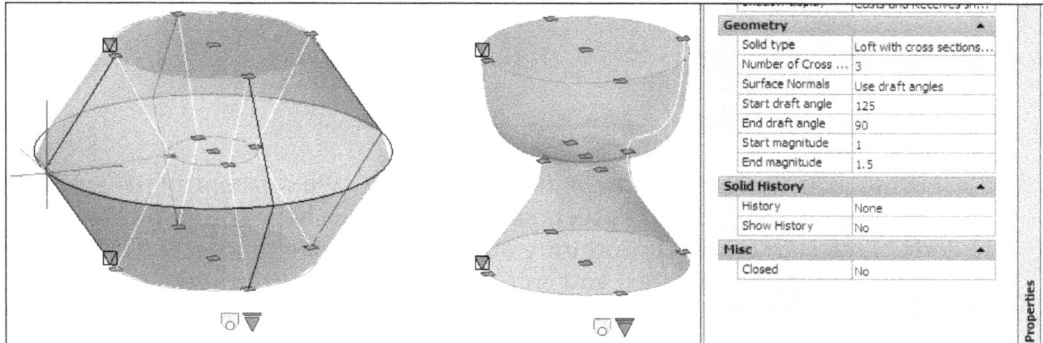

Some more options are available:

- **POint**: This option prompts us to specify a point that initiates or ends the loft object, without needing a point object. All other sections must be closed.

- **Join multiple edges**: This option allows us to select multiple continuous objects as one single section.

- **Guides**: This option allows us to select one or more guided curves. Besides specifying all cross sections, we can further model the lofted object.

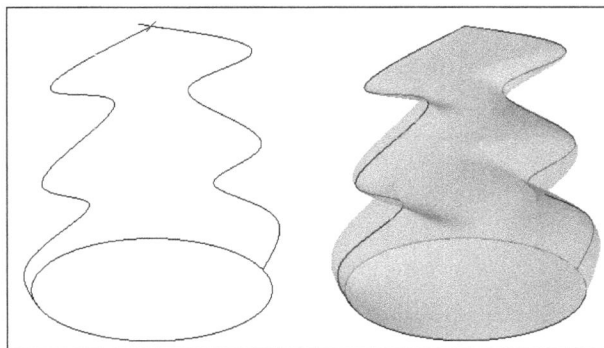

- **Path**: This option allows us to select a linear object that acts as a path. This object must intersect the planes of all sections.

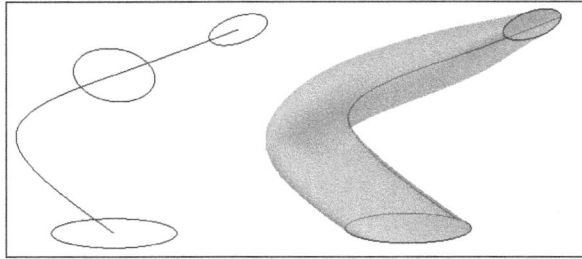

- **Settings**: This option displays a dialog box where we can control the surface of the created image at cross sections. If the chosen shape is **Smooth Fit**, we control the start and end continuity and the bulge magnitudes; if it is **Draft Angles**, we control the start and end angles and the magnitudes.

Exercise 4.4

We are going to create a bathtub from scratch, with the necessary cross sections and guides, by following these steps:

1. Create a rectangle of 2.2 x 1.2 units.
2. Create another rectangle of 1.8 x 0.9 units with a fillet radius of 0.2 in all corners and center its width with the first rectangle.
3. Copy both objects by 0.6 units in the Z direction.
4. The inner rectangle at the bottom is moved by 0.15 in the Z direction.
5. The rectangle that is selected last is scaled by 0.92 units to a point near the right-hand edge.

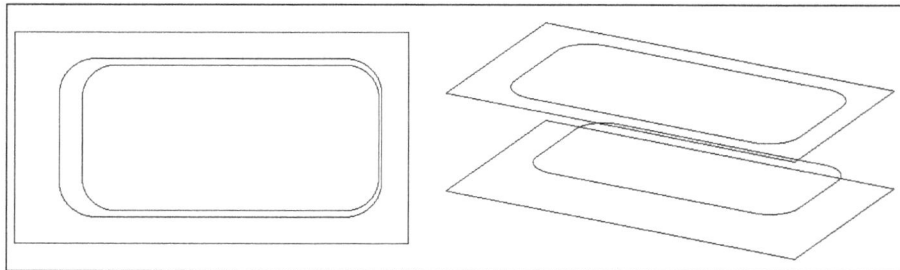

6. Just to test, apply LOFT without guides. Then create and activate a layer called BATHTUB with a different color.

7. If we want to undo the previous steps, apply the UNDO command (type UNDO to access this command; do not confuse this with the U command) and select the **Mark** option.

8. Applying the LOFT command, select, in this order: the bottom rectangle, top rectangle, top-inner rectangle, and bottom-inner rectangle. Pressing the *Enter* key twice will get us an interesting object, but not what we want. Even trying other loft shapes, the result is not famous.

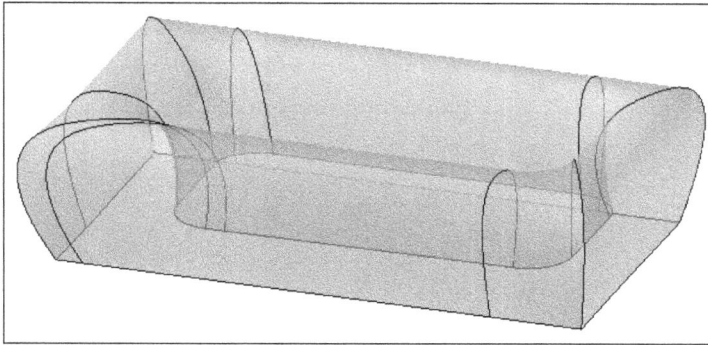

9. Apply the UNDO command and select the **Back** option to cancel the LOFT function.

10. With the current layer set to 0, draw four guides that will mold the loft. Create a UCS that is vertically aligned at the middle of length and draw a spline with the wanted section that intersects all four sections. Then mirror once and rotate with copy to obtain all four splines. Adjust all vertices with the adequate UCS.

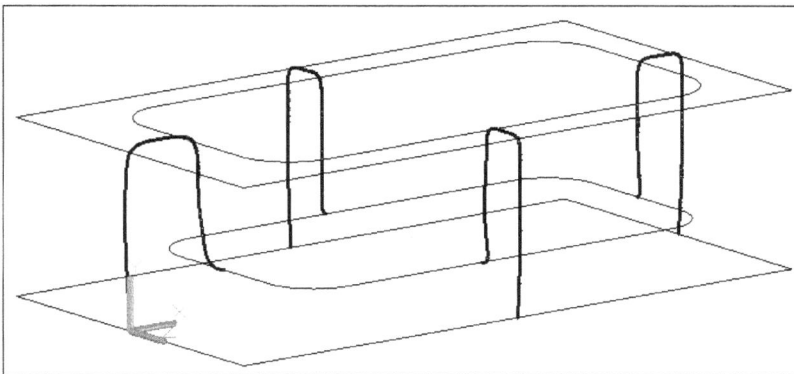

[We have to apply splines as guides. The 2D polylines can only be
applied if composed by only one segment. The easiest method to adjust
vertices is to turn off the **Osnap** function and apply the SPLINEDIT
command and select the **Edit vertex** option.]

11. Finally, activate the BATHTUB layer and apply the LOFT command; first select
    the four cross sections in the same order, press the *Enter* key, and choose the
    Guides option, and then select each of the four splines. As we select each
    spline, the lofted object adjusts accordingly. Then press the *Enter* key to exit
    the command.

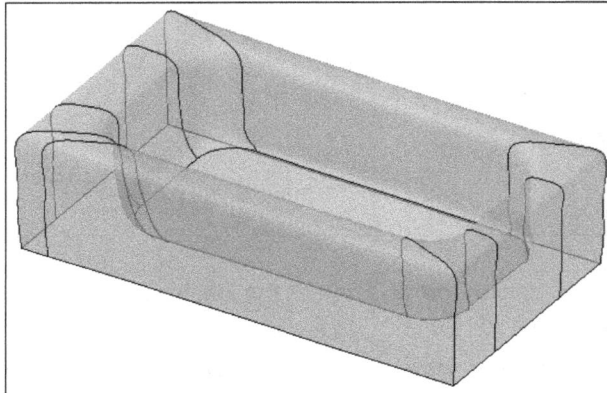

12. Save the model with the name A3D_04_04FINAL.DWG.

Creating solids and surfaces along a path

The last command to create solids and surfaces from linear objects is presented next.

The SWEEP command

The SWEEP command (no alias) creates a solid or surface by extruding sections
along a path. If the section is open or if the **MOde** option is set to **Surface**, a surface
is created.

The path can be a line, arc, elliptical arc, 2D or 3D polylines, 2D or 3D splines, circles,
ellipses, helixes, and surface or solid edges.

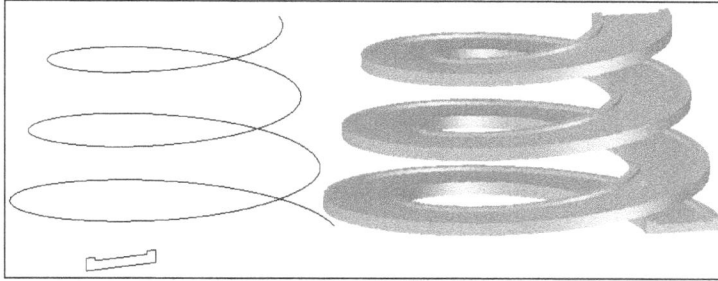

This command starts by displaying the same information as in previous commands and asks for one or more objects that define the section:

```
Command: SWEEP
Current wire frame density:  ISOLINES=4, Closed profiles creation mode
= Solid
Select objects to sweep or [MOde]: Selection
```

By default, we select the linear object that defines the path, and the sweep object is created. The section is placed normal to the path with its center aligned to it:

```
Select sweep path or [Alignment/Base point/Scale/Twist]: Selection
```

We have some options available:

- **Alignment**: This option specifies if the section is rotated to be perpendicular to the path or if its plane is maintained

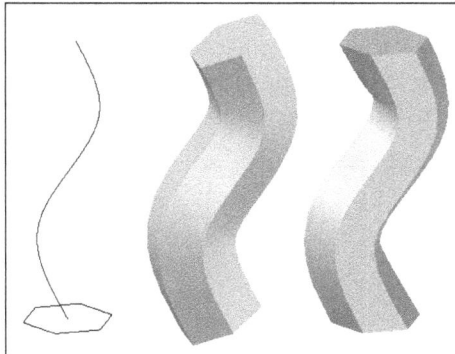

- **Base point**: This option prompts for the point of the section that will follow the path
- **Scale**: This option enables us to specify a scale factor for the section at the end of the path

- **Twist**: This option enables us to specify a rotation angle for the section at the end of the path

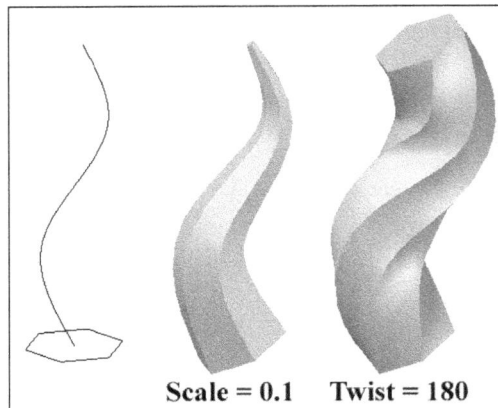

Scale = 0.1   Twist = 180

> Any object selected as a section should be oriented with the XY plane. If the object obtained with the command gets a reversed section, we undo the command, apply the MIRROR command to the section, and then repeat SWEEP. Another possibility with open paths is to select the path near the opposite endpoint.

Exercise 4.5

With the SWEEP command, we are going to create an ogee on the line that represents a ceiling by following these steps:

1. Create the following 2D polylines (the one on left-hand side will be the path and the one on right-hand side will be the section):

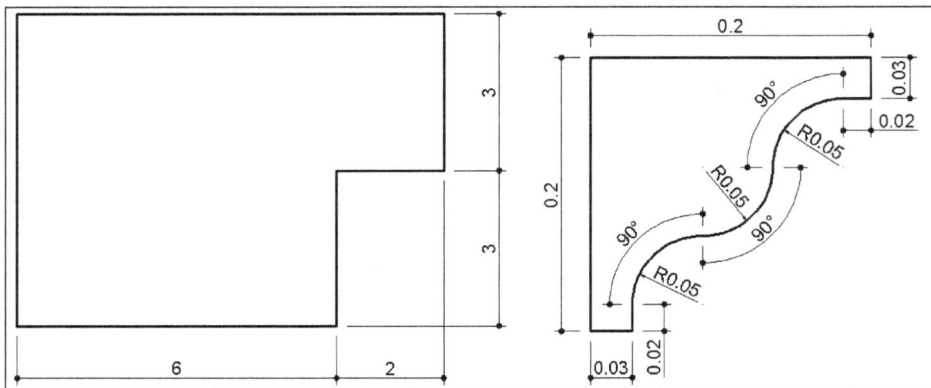

2. With the LAYER command, a layer for the sweep object is created and activated.

3. Now apply the SWEEP command. The polyline on the right-hand side is the section. Before the path, choose the **Base point** option and specify the upper-left corner of the section as the point that will follow the path. Finally, select the polyline on left-hand side as the path.

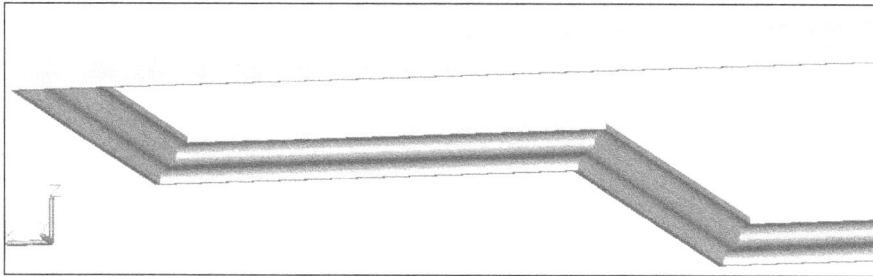

4. Save the model with the name A3D_04_05FINAL.DWG.

> If the section is not applied correctly, that is, it is upside down or with the worked part out, undo the command and apply the MIRROR command. Then apply SWEEP.

Exercise 4.6

The piping systems can be easily created with the SWEEP command. To do so, follow these steps:

1. Open the drawing A3D_04_06.DWG. This drawing already contains a 3D polyline that will be used to create a set of pipes like those used in industrial facilities.

2. To define the section, create three circles, set the radius as `0.5` and separated them by `0.1` units in the Y direction.

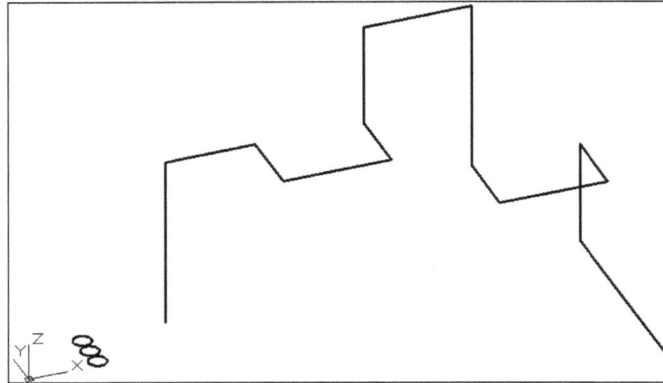

3. Apply the `SWEEP` command and select the three circles as a section. Choose the **Base point** option and specify the center of the top circle. Finally, select the 3D polyline as the path.

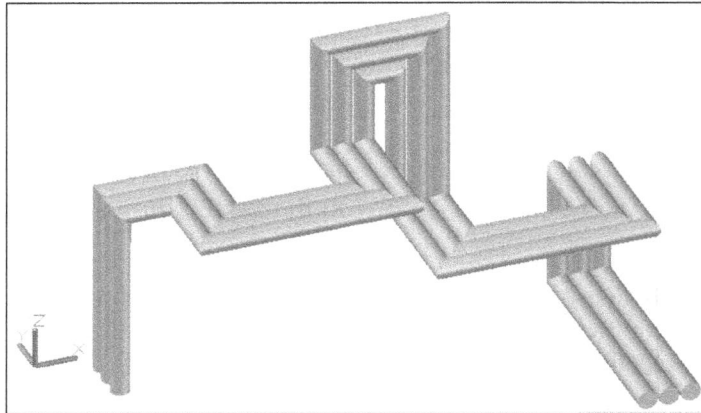

4. Save the model with the name `A3D_04_06FINAL.DWG`.

Important variables for solids and surfaces

As in 2D, AutoCAD has also some variables responsible for the creation or visualization of 3D objects:

- `ISOLINES`: This variable controls the number of lines used to represent curved faces of solids with the wireframe visual style. These isolines should not be confused with edges. By default, its initial value is 4 and it is saved with the drawing.

- `DISPSILH`: This variable controls the display of silhouette edges of solids when applying a wireframe visual style. It is saved with the drawing and accepts two values, 0 (the silhouette is not displayed; this is the default) and 1 (silhouettes are displayed).

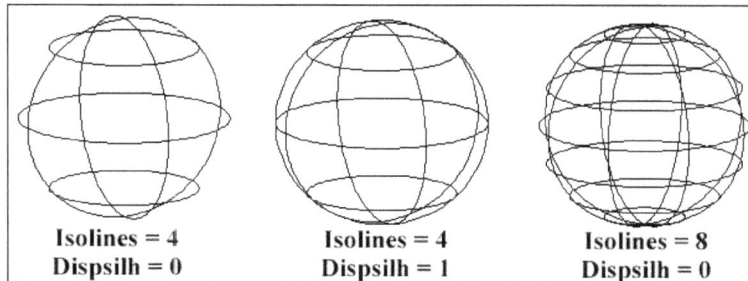

| Isolines = 4 | Isolines = 4 | Isolines = 8 |
| Dispsilh = 0 | Dispsilh = 1 | Dispsilh = 0 |

- `FACETRES`: This variable controls the amount of faceted faces used to represent shaded or rendered objects. All shaded or rendered views are calculated by faceting all curved surfaces and faces. By default, its initial value is 0.5 and it is saved with the drawing. Its value can go from 0.01 to 10; a reasonable value is 2 or 3.

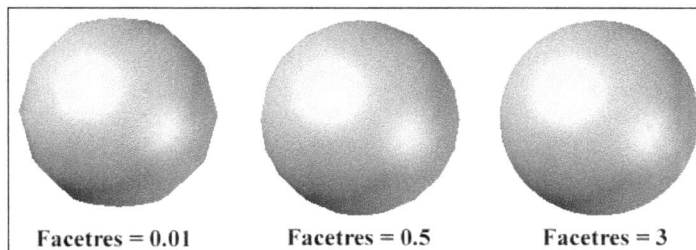

| Facetres = 0.01 | Facetres = 0.5 | Facetres = 3 |

- `DELOBJ`: This variable controls if linear objects used to create solids or surfaces are deleted. By default, its initial value is 3 and it is saved in the registry. It accepts the following values: 0 (all geometry is maintained), 1 (deletes only section objects), 2 (deletes sections, paths, and guide objects), and 3 (same as 2, but only if a solid is created). It can also accept -1, -2, and -3 with similar behavior as positive values, but requests confirmation for deletion.

- SURFACEASSOCIATIVITY: This variable controls if surfaces maintain an association to original linear objects. By default, its initial value is 1 and it is saved with the drawing. It accepts two values, 0 (association is not maintained) and 1 (association is maintained). With a value of 1, the DELOBJ variable is ignored when creating surfaces.

- SURFU and SURFV: These variables control the density of the surface isolines in the first and second directions. By default, their initial value is 6 and they are saved with the drawing.

- LOFTPARAM: This variable controls the shape of lofted objects. By default, its value is 7. It is saved with the drawing and is defined by the following values: 1 (sections do not twist), 2 (aligns directions in each section), 4 (simplifies solids and surfaces), and 8 (closes the lofted object between the first and last sections).

- LOFTNORMALS: This variable controls the shape of the lofted object in terms of transition between sections. By default, its initial value is 1 and it is saved with the drawing. It accepts the following values: 0 (ruled), 1 (smooth fit), 2 (normal to the first section), 3 (normal to the last section), 4 (normal to the first and last sections), 5 (normal to all sections), and 6 (apply the draft angle and magnitude).

- LOFTANG1 and LOFTANG2: These variables define the tangency angles at the first and last sections of the lofted object. By default, their initial values are 9 and 0 and they are saved with the drawing.

- LOFTMAG1 and LOFTMAG2: These variables define the magnitude (intensity) of draft angles at the first and last sections of the lofted object. By default, their initial values are 0 and they are saved with the drawing.

Summary

In this chapter, we learned the most important concepts and commands to create solids and surfaces from 2D objects. We learned about the EXTRUDE and PRESSPULL commands for extrusions (giving height), REVOLVE for revolving objects, LOFT for objects defined by two or more cross sections, and SWEEP to create objects by extruding along a path. These commands provide the foundation for 3D modeling. Next, we will see basic 3D shapes and object conversions.

5

3D Primitives and Conversions

In the previous chapter, we saw the creation of solids and surfaces based on linear objects. In the current chapter, we will see the remaining commands to create 3D solids and 3D surfaces, followed by conversion commands.

The topics covered in this chapter are as follows:

- Creating 3D solid basic shapes
- Creating planar surfaces
- Creating planar surfaces delimited by linear objects
- How to convert objects to solids
- How to convert objects to surfaces

3D solid primitives

The 3D primitives are solid basic shapes like boxes, cylinders, spheres, and others. We can also apply a special solid primitive to create walls in a similar way as we create a polyline.

Accessing all these commands, besides digitizing their names or alias, can be done on the ribbon, **Home | Modeling** panel, on the **Modeling** toolbar, and on the **Draw | Modeling** Menu bar:

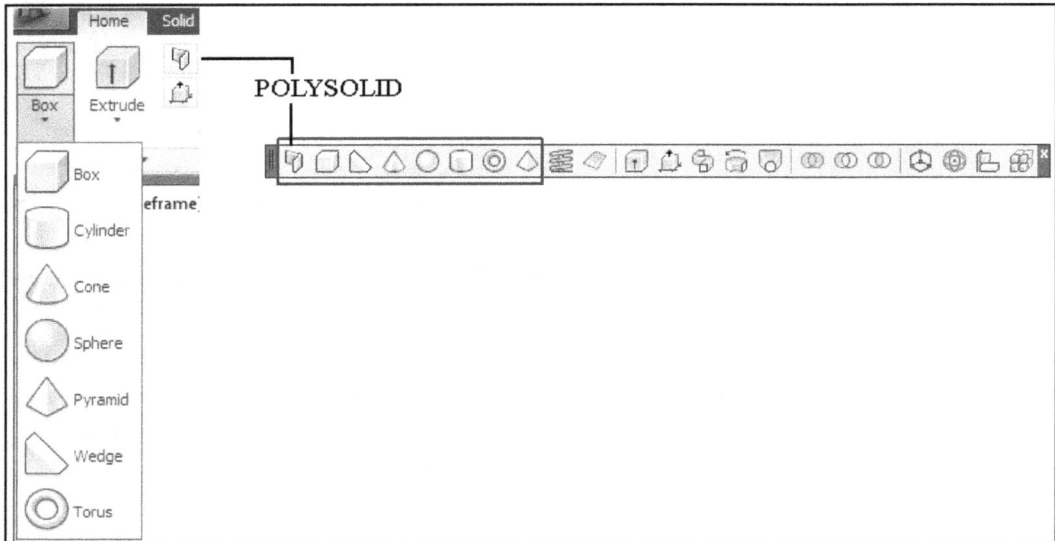

The creation of these solid primitives, like most commands in the previous chapter, is dynamic, that is, upon specifying one or more points or dimensions, we see the objects being formed.

> For most basic shapes (with the exception of POLYSOLID), we can use the **PROPERTIES** palette (*Ctrl + 1*) to change dimension parameters after creating the solid.

To improve the visibility of curved faces in non-wireframe visual styles, we may apply the FACETRES variable. A value of 2 or 3 is normally enough. With wireframe visual styles, we may apply the ISOLINES variable to increment the number of lines used to represent curved faces, or the DISPSILH variable to show a silhouette or profile lines of solids with curved faces. These variables were explained at the end of *Chapter 4, Creating Solids and Surfaces from 2D*.

Creating boxes and wedges

We start with the most basic 3D shape, the box. Wedges are boxes cut in two; dialog and options are exactly the same.

The BOX command

The BOX command creates boxes, that is, a solid with six rectangular and orthogonal faces, aligned with the axes of the current user coordinate system.

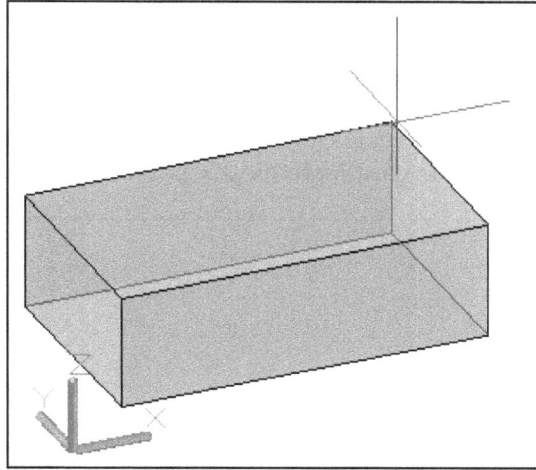

By default, the command asks for the first corner and the opposite corner:

```
Command: BOX
Specify first corner or [Center]: Point
Specify other corner or [Cube/Length]: Point
```

These two points must not be on the same axis; at least they must define an area. If these two points are on the same plane, the height is requested. If we specify a negative value, the box is created downward:

```
Specify height or [2Point]: Point or value
```

> The easiest way to create a box is to specify the first corner and define the opposite corner as relative coordinates in the three directions, like @length, width, height.

Options for this command:

- **Center**: Instead of specifying the first corner, we can start by defining the center of the box. Then, we define the other corner or options.
- **Cube**: After defining the first corner or center, this option allows for creating a cube and requests the cube length. If we pick a point, the cube may be rotated.

- **Length**: After defining the first corner or center, this option allows you to specify the length (along the X axis), width (along the Y axis), and height (along the Z axis) of the box, in this order. If we pick a point to specify the length, the box may be rotated.

- **2Point**: The height of the box can be defined by specifying two points.

The WEDGE command

The WEDGE command (alias WE) creates a wedge- a box cut by a diagonal plane. The triangular faces stay on a plane parallel to the ZX plane.

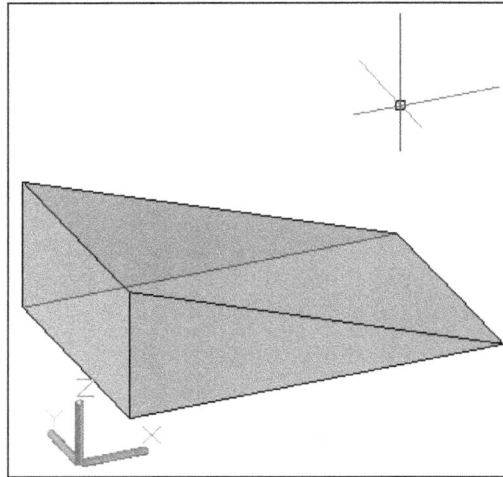

The dialog is equal to the BOX command; by default, we specify two corners:

```
Command: WEDGE
Specify first corner or [Center]: Point
Specify other corner or [Cube/Length]: Point
```

If these points are on the same plane, the height is requested. The inclined face points toward the X axis:

```
Specify height or [2Point]: Point or value
```

Options for this command:

- **Center**: Instead of specifying the first corner, we can start by defining the center of the wedge. Then, we define the other corner or options.

- **Cube**: After defining the first corner or center, this option allows you to create a half cube cut by a diagonal plane and requests the cube length.

- **Length**: After defining the first corner or center, this option allows you to specify length (along the X axis), width (along the Y axis), and height (along the Z axis) of the wedge, in this order.

- **2Point**: The height of the wedge can be defined by specifying two points.

Exercise 5.1

To practice with these objects, we are going to create the basic shapes to construct a building similar to a cathedral.

1. We start a new drawing and create a layer called 3D-CATHEDRAL; apply the green color and make it Current.

2. We create a box with length 100, width 40, and height 45. The first corner can be at the origin and the opposite corner at 100,40,45.

3. We create another box whose center is at the middle of the top face. We specify the **Center** option and use the Otrack function to find the center of the top face. Another way to find the center is to apply individually **3D Osnap Center of face** (*Ctrl* and mouse right button). Then the opposite corner is @15,15,20. Later, we will combine these boxes.

4. Again with BOX, we create one tower with the dimensions 10,10,30 at one of the top corners.

5. We copy the last box to the other's front top corner.

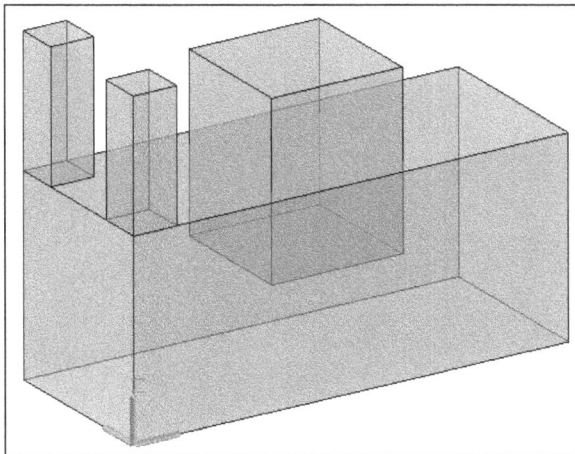

6. Now it's time to create two wedges to act as some kind of buttress. We apply WEDGE, first at the lower-left corner of the box and the dimensions 20,5,40.

7. We create another wedge, same first corner but with dimensions 15,5,30.

8. As both wedges oriented along X, we apply the ROTATE command to rotate them around the first corner -90 degrees. Later we will subtract the second to the first wedge.

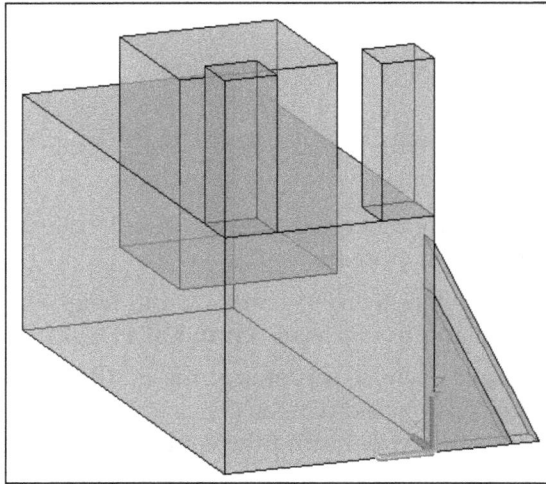

9. We copy these two wedges twice to have one set at the middle of the edge and the other at the end.

10. Finally, we mirror these six wedges to the other side of the building.

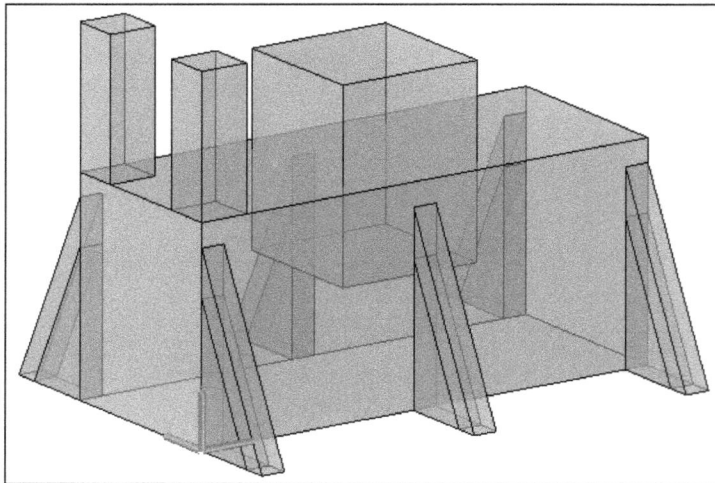

11. Later we will continue. For now, save the drawing with the name A3D_05_01FINAL.DWG.

Creating cylinders and cones

The next two 3D basic shapes that we are going to see are cylinders and cones.

The CYLINDER command

The CYLINDER command (alias CYL), creates cylinders with a circular or elliptical base, that is, a solid with two planar circular or elliptical faces and one curved face.

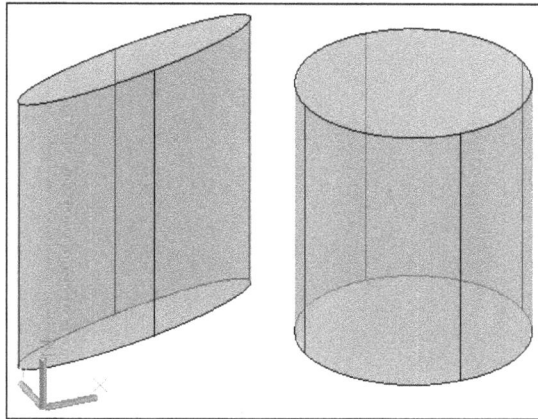

By default, the command starts by prompting the center of the base and the radius of the base:

```
Command: CYLINDER
Specify center point of base or [3P/2P/Ttr/Elliptical]: Point
Specify base radius or [Diameter]: Value or point
```

Next, we have to specify the height of the cylinder:

```
Specify height or [2Point/Axis endpoint]: Value or point
```

The CYLINDER command has the following options:

- **3P/2P/Ttr**: These options are identical to the options of the CIRCLE command, thus allowing us to define the base circle by three points, two points on diameter or tangent to two existing objects, and a specified radius respectively.
- **Elliptical**: This allows us to define an elliptical base for the cylinder with the same options of the ELLIPSE command. By default, it prompts for two points that define the first axis and the distance of the second semiaxis. The **Center** option defines the elliptical base starting from the center.

- **Diameter**: Instead of defining the radius of the base, this option allows us to specify the diameter.
- **2Point**: It allows us to define the height of the cylinder by two points.
- **Axis endpoint**: This option allows you to specify the position of the center of the top face. If the cylinder axis (defined by the two centers) is not parallel to the Z axis of the current user coordinate system, both planar faces are rotated so that they stay perpendicular to the cylinder axis.

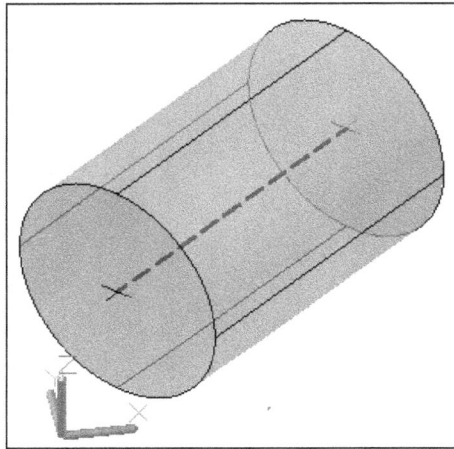

The **Axis endpoint** option allows us to create cylinders with bases not aligned to the current XY working plane. This may avoid the creation of specific user coordinate systems.

The CONE command

The **CONE** command (no alias) allows you to create cones or frustum cones. In the first case, the object is composed by a curved face and a planar face. In the second case, there are two parallel planar faces and a curved face.

By default, the command prompts for the center of the base and radius:

```
Command: CONE
Specify center point of base or [3P/2P/Ttr/Elliptical]: Point
Specify base radius or [Diameter]: Value or point
```

Next, we have to specify the height of the cone:

```
Specify height or [2Point/Axis endpoint/Top radius]: Value or point
```

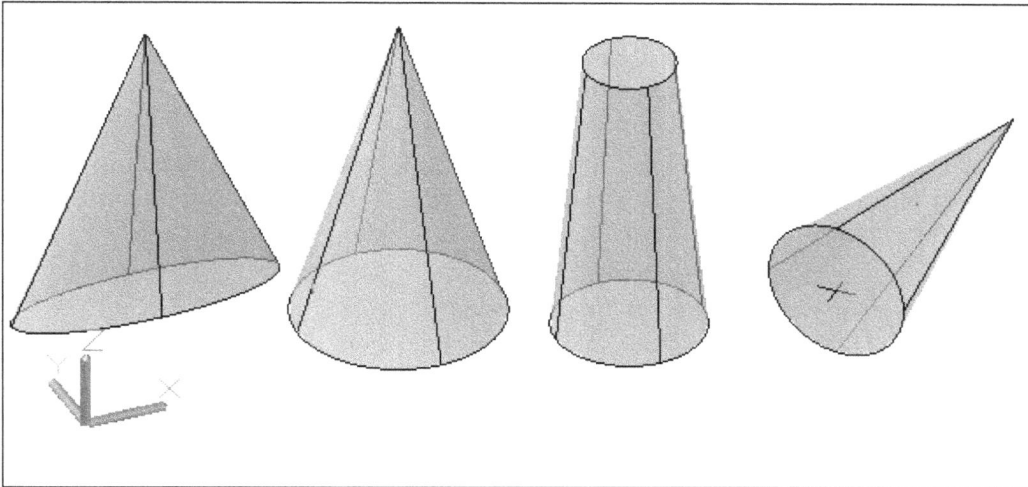

Most of the options are similar to those of the CYLINDER command. They are as follows:

- **3P/2P/Ttr**: These options are identical to the options of the CIRCLE command, thus allowing us to define the base circle by three points, two points on the diameter or tangent to two existing objects, and a specified radius respectively.

- **Elliptical**: This allows us to define an elliptical base for the cone with the same options as the ELLIPSE command. By default, it prompts for two points that define the first axis and the distance of the second semiaxis. The **Center** option defines the elliptical base starting from the center.

- **Diameter**: Instead of defining the radius of the base, this option allows us to specify the diameter.

- **2Point**: This allows us to define the height of the cone by two points.

- **Axis endpoint**: This option allows us to specify the position of the top vertex of the cone. If the cone axis is not parallel to the Z axis of the current user coordinate system, the cone is rotated.

- **Top radius**: This option prompts for the radius of the top face. If not zero, it creates a frustum cone.

Exercise 5.2

Now, we continue exercise 5.1:

1. Let's open the drawing saved in *exercise 5.1* or `A3D_05_02.DWG`.

2. We create first the cylinder with its center of base at the middle of the top face of the higher box (using `Otrack` from the midpoint or **3D Osnap Center of face**) with a radius of `13` and a height of `2`.

3. Continuing to model that part of the building, we create a new cylinder with its center of base at the center of the top face of the previous cylinder, with a radius of `8` and a height of `10`.

4. To create the columns, again we use the cylinder with its center at a 180 degree quadrant of the top face of the first cylinder, with a radius of `0.8` and a height of `10`.

5. We move this cylinder inwards by `1`, that is, in the X direction.

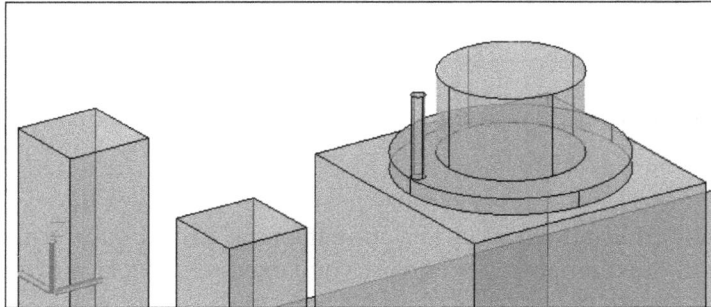

6. We apply the `ARRAY` command, choose the **Polar** option, the center of the array is the center of the first cylinder, and define `16` items filling an angle of 360 degrees.

7. We copy the first cylinder `12` meters along the Z direction.

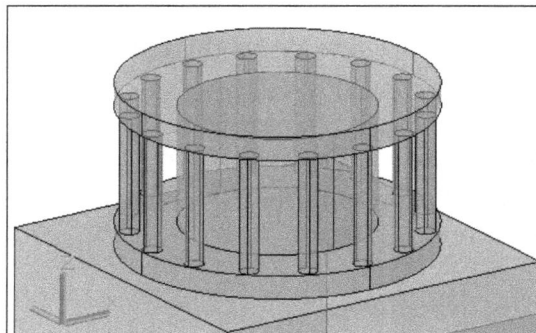

8. On the back of the building, we create another cylinder with the center at the midpoint of lower edge having a radius of `10` and a height of `30`.

9. On top of this cylinder, we create a cone with the same radius and height, `15`.

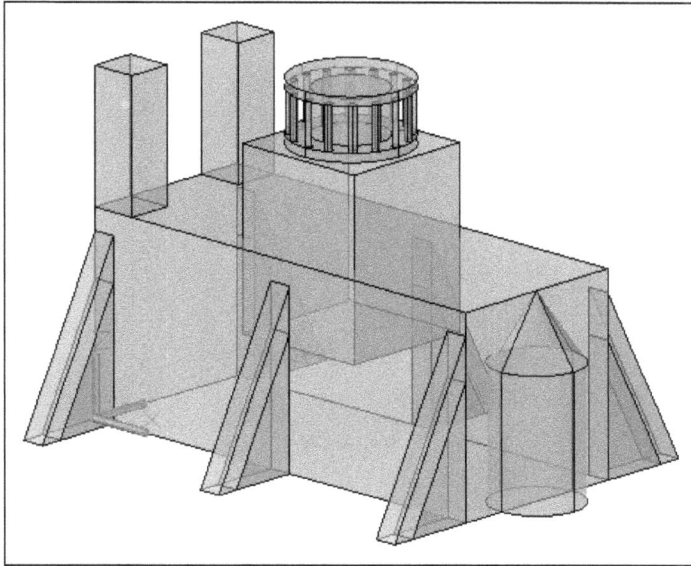

10. We save the drawing with the name `A3D_05_02FINAL.DWG`.

Creating spheres

Another basic shape is the sphere or globe, created with the next command.

The SPHERE command

The `SPHERE` command (no alias) creates spheres. This 3D shape has only one curved face; all points on the face are at the same distance from the center. It has no edges or vertices.

By default, the command prompts only for the center of the sphere and radius:

```
Command: SPHERE
Specify center point or [3P/2P/Ttr]: Point
Specify radius or [Diameter]: Value or point
```

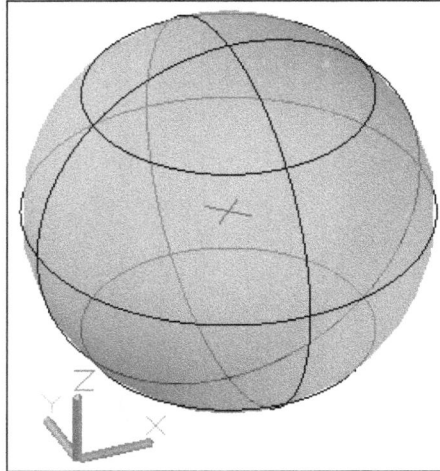

The options are also similar to those of the previous commands:

- **3P/2P/Ttr**: We may define the circumference of the sphere by three points, two points on diameter or tangent to two existing objects, and a specified radius

- **Diameter**: Instead of defining the radius of the sphere, this option allows us to specify the diameter

> With a wireframe visual style, spheres are shown only with two meridian and three parallel isolines. We may increase the value of the ISOLINES variable. With shaded visual styles or in render, we get better results by increasing the value of the FACETRES variable.

Creating 3D donuts

If we want a solid shaped like a donut or torus, the next command is a possibility. This shape can also be obtained by applying the REVOLVE command to a circle.

The TORUS command

The TORUS command (alias TOR) creates toroid solids, that is, solids with a 3D donut shape.

By default, the command prompts for the center of the torus, the radius of torus, and the radius of the circular section. The torus bisects the XY plane of the current user coordinate system:

```
Command:  TORUS
Specify center point or [3P/2P/Ttr]: Point
Specify radius or [Diameter]: Value or point
Specify tube radius or [2Point/Diameter]: Value or point
```

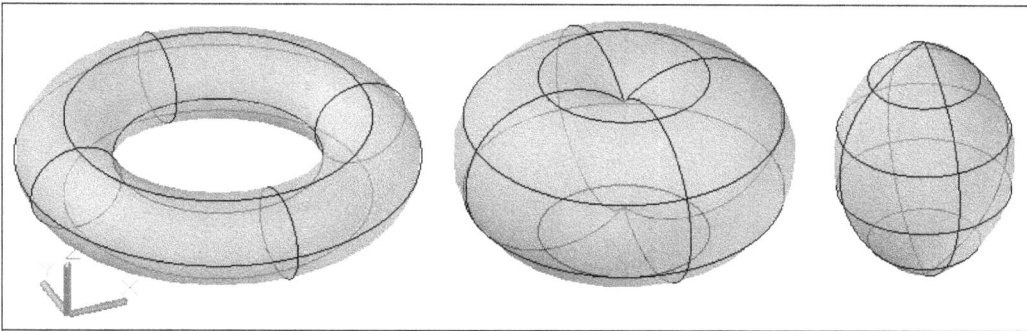

Besides the 3D donut shape, where the torus radius is greater than the section radius, this solid primitive can assume two particular shapes:

- **Apple shape**: The torus radius is smaller than the section radius
- **Rugby or American football ball shape**: The torus radius is negative and with the absolute value smaller than the section radius

The options for this command are:

- **3P/2P/Ttr**: We may define the circumference of the torus by three points, two points on diameter or tangent to two existing objects, and a specified radius
- **Diameter**: Instead of defining the radius of the torus, this option allows us to specify the diameter
- **2Point**: The radius of the circle's section can be defined by two points

Exercise 5.3

We are placing some more elements into our cathedral.

1. Open the drawing saved in *exercise 5.2* or A3D_05_03.DWG.

2. Create a sphere on top of the highest cylinder, with its center coincident with the center of the top face and radius 13 (this can also be reached by a quadrant point). Later we will cut the sphere in half, forming a dome.

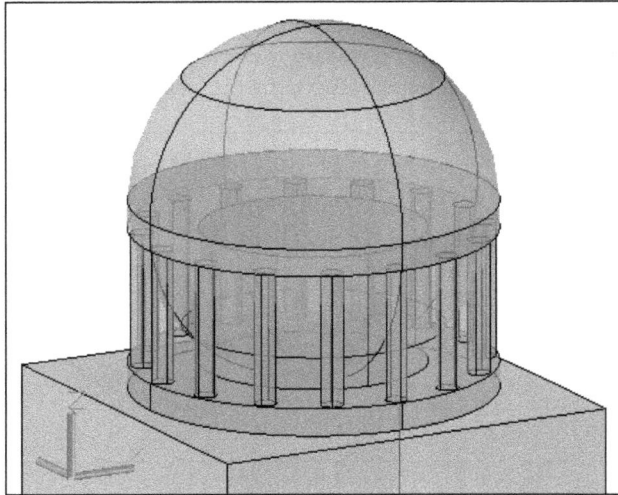

3. With the UCS command, create a user coordinate system aligned with the front facade. The origin is the lower-left corner of the first box, the X axis defined by the other lower corner, and the XY plan defined by one of the top corners of that face. This UCS also allows us to control the precise position of the torus.

4. Applying the TORUS command, pause over the midpoint on the lower edge, point along Y, and write 30. The Otrack function must be on. Another way is to specify the torus origin at 20,30 absolute coordinates. The radius of the torus is 11 and the radius of the section is 1.5.

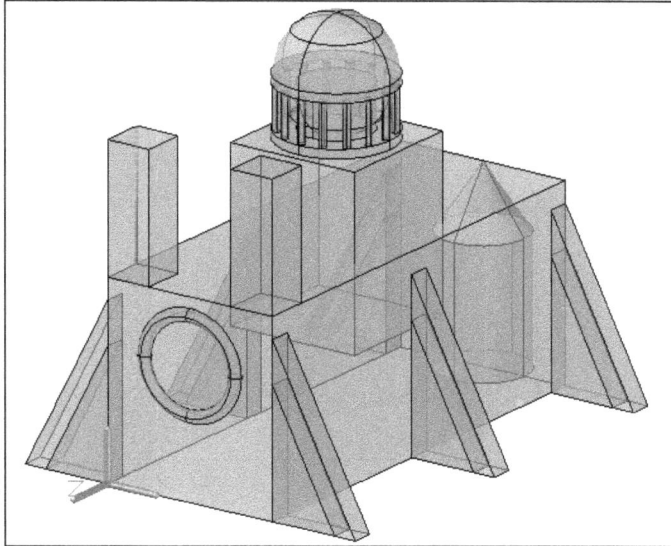

5. Activate the world coordinate system (for instance, with the default option of the UCS command).

6. Save the drawing with the name A3D_05_03FINAL.DWG.

Creating pyramids

Next is a command to create pyramids and pyramid frustums with regular bases.

The PYRAMID command

The PYRAMID command (alias PYR) allows you to create pyramids where the base is a regular polygon with sides of equal size. Also, it is possible to create pyramid frustums.

By default, the command displays information about the current settings of the base and prompts for the center of the base:

```
Command: PYRAMID
4 sides  Circumscribed
Specify center point of base or [Edge/Sides]: Point
```

Next we specify the radius of the regular polygon that defines the base of the pyramid, according to the displayed option:

```
Specify base radius or [Inscribed]: Value or point
```

Finally, we specify the height of the pyramid:

```
Specify height or [2Point/Axis endpoint/Top radius]: Value or point
```

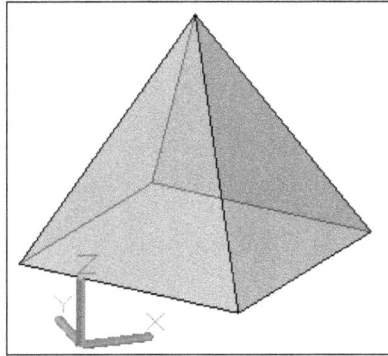

The options available for this command are:

- **Edge**: Instead of center and radius, we specify two points that define one of the sides for the base of the pyramid, similar to the option of the POLYGON command.

- **Sides**: This allows us to change the number of sides of the polygon that define the base of the pyramid.

- **Inscribed/Circumscribed**: The polygon that defines the base of the pyramid can be inscribed (the radius defines the distance to vertices), or circumscribed (the radius defines the distance to midpoints).

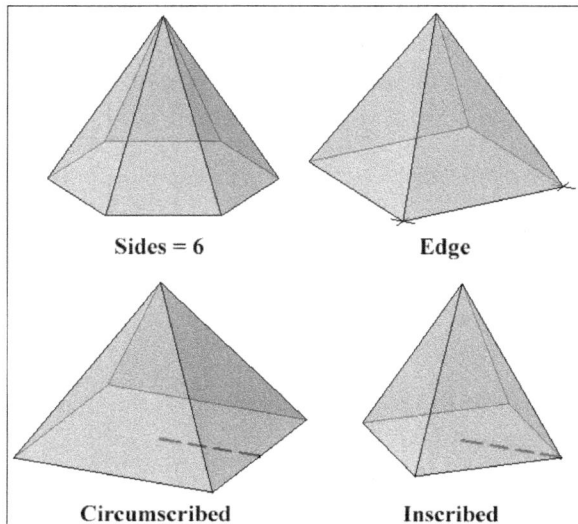

Sides = 6   Edge

Circumscribed   Inscribed

- **2Point**: This allows us to define the height of the pyramid by two points.

- **Axis endpoint**: This option allows us to specify the position of the top vertex of the pyramid. If the pyramid axis is not parallel to the Z axis of the current user coordinate system, the pyramid is rotated.

- **Top radius**: This option prompts for the radius of the top face. If not zero, it creates a pyramid frustum.

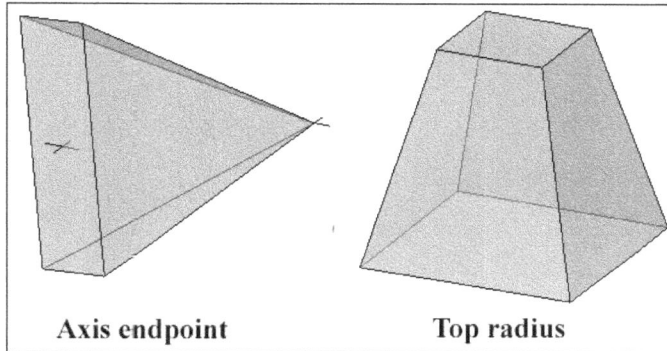

Axis endpoint   **Top radius**

Creating walls

The last 3D solid primitive allows us to create some kind of walls in a different way from the EXTRUDE or PRESSPULL commands.

The POLYSOLID command

The POLYSOLID command (or PSOLID, but too long to be an alias) allows us to create solids similar to walls. What this command really does is it takes a wall section and applies a sweep to it along a path defined by the user. The path can have several segments or arcs and its construction is similar to drawing a 2D polyline.

> This is the only primitive that we can't change with the **PROPERTIES** palette; the object is recognized as a swept solid, as if it was created by the SWEEP command.

By default, the command displays information about the height, width, and justification, and just prompts for points:

```
Command: POLYSOLID
Height = 80.0000, Width = 5.0000, Justification = Center
Specify start point or [Object/Height/Width/Justify] <Object>: Point
Specify next point or [Arc/Undo]: Point
Specify next point or [Arc/Undo]: Point
```

If we want to draw a wall along an arc, we apply the **Arc** option:

```
Specify next point or [Arc/Close/Undo]: Arc
Specify endpoint of arc or [Close/Direction/Line/Second
point/Undo]: Point
```

To return to segments, we apply the **Line** option:

```
Specify endpoint of arc or [Close/Direction/Line/Second point/Undo]:
Line
Specify next point or [Arc/Close/Undo]: Point
Specify next point or [Arc/Close/Undo]: Point
Specify next point or [Arc/Close/Undo]: Point
Specify next point or [Arc/Close/Undo]: Point
```

To end the command, we just press *Enter*:

```
Specify next point or [Arc/Close/Undo]: Enter
```

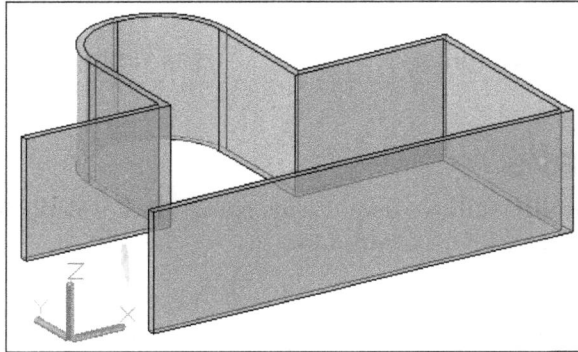

This command has the following options:

- **Object**: This option prompts for the selection of one linear object and applies this object as the path. The DELOBJ variable controls if the object is erased. Eligible objects are lines, arcs, circles, and 2D polylines.

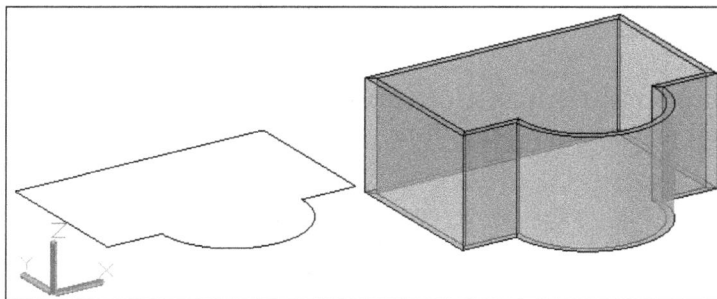

- **Height**: This option controls the height of the wall section.

- **Width**: This option controls the width of the wall section.

- **Justify**: This option accepts left, center, or right justification, and controls if the points we specify to define the path correspond to the left face, center of the wall, or right face.

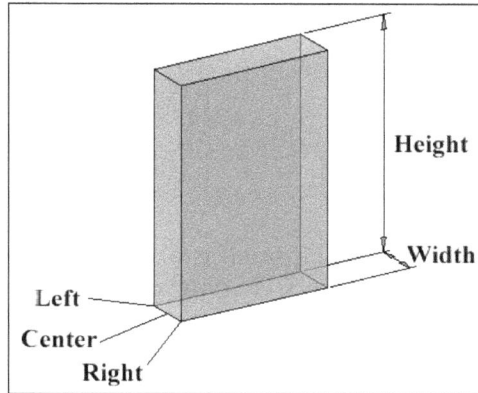

- **Close/Undo**: When specifying points, the **Close** option closes the wall and the **Undo** option cancels the last point, like the PLINE command.

- **Arc/Line**: When specifying points, **Arc** changes to mode Arc to add arcs to the path. By default, arcs are continuous with the additional options **Direction** and **Second Point**, allowing directional changes. **Line** changes back to mode line.

> For complex paths, it is better to draw a 2D polyline and then apply the **Object** option.

Exercise 5.4

We continue the exercise with the application of pyramids and a wall by performing the following steps:

1. Open the drawing saved in *exercise 5.3* or A3D_05_04.DWG.

2. On top of one of the towers, apply the PYRAMID command, the **Edge** option, and specifying two side corners with the height 15.

3. Copy the pyramid to the other tower.

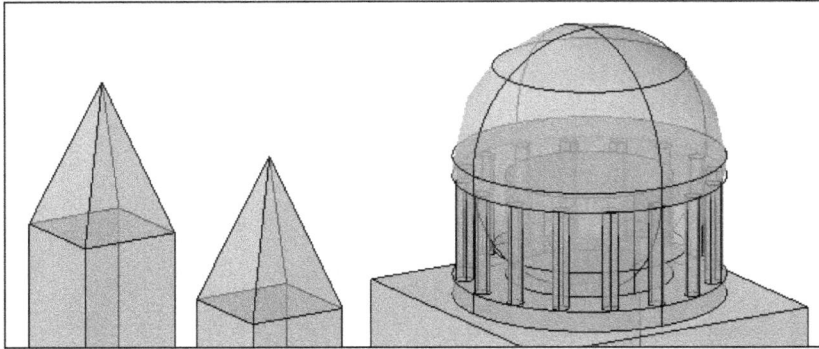

4. Finally, we are going to build a small wall on the top of the largest box. We apply the POLYSOLID command, change height to 1.2 units, width to 0.2 units, and *justify* it to the *right*. Then we specify the points counterclockwise.

5. Save the drawing with the name A3D_05_04FINAL.DWG.

Primitive surfaces

In the previous chapter, we explained how to create surfaces from linear objects with the EXTRUDE, REVOLVE, SWEEP, or LOFT commands. Here we present a command to create rectangular surfaces or delimited by planar objects.

Creating planar surfaces

Here is a command to create planar surfaces.

The PLANESURF command

The PLANESURF command (no alias) creates planar surfaces defined by two opposite corners or delimited by one or more existing objects.

By default, the command just prompts for two opposite corners and creates a rectangular surface:

```
Command: PLANESURF
Specify first corner or [Object] <Object>: Point
Specify other corner: Point
```

The command has only one option, **Object**.

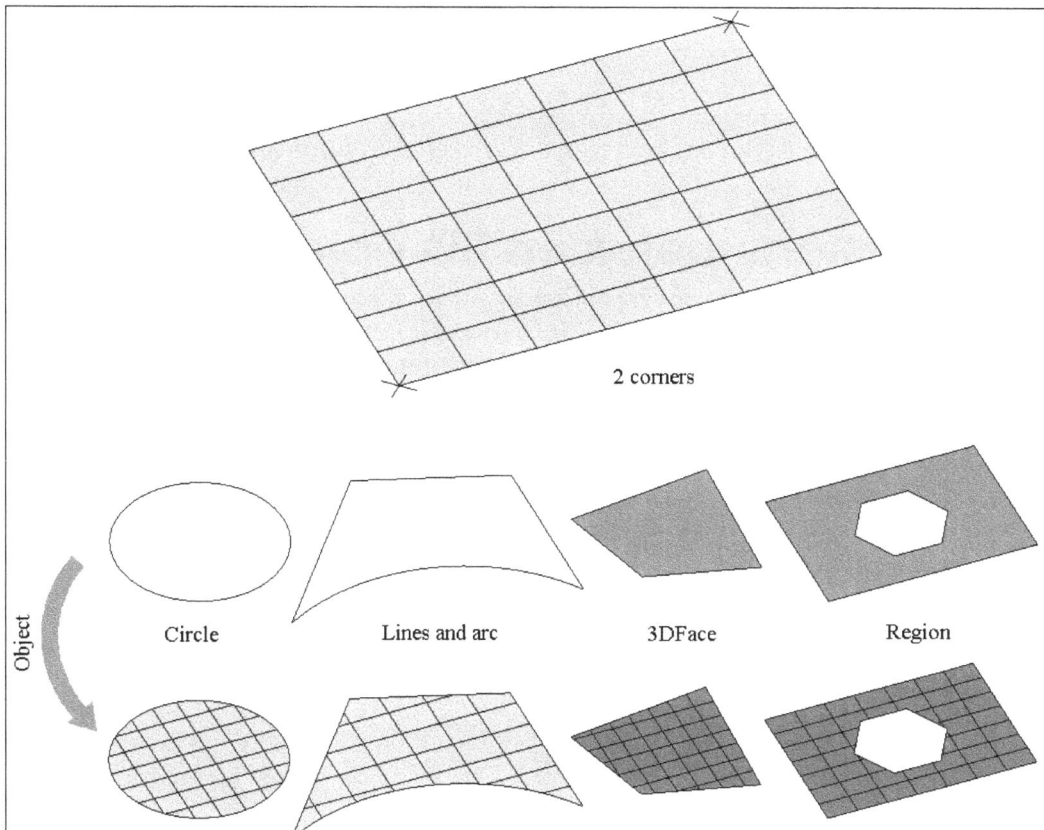

- **Object**: This option prompts for the selection of objects. We can select lines, arcs, circles, ellipses, 2D polylines, 3D planar polylines, planar splines, regions, and 3D faces. For each closed object or set of objects that form a closed boundary, the option creates a surface. By default, the original objects are maintained in the drawing.

> There are two important variables when creating surfaces from other objects. SURFACEASSOCIATIVITY controls if the surface remains connected to the original objects (value 1) or not (value 0). This variable is saved with the drawing. With value 1, the DELOBJ variable, saved with AutoCAD, is not applied to surfaces and all original objects are not deleted. To delete the original objects when creating surfaces, SURFACEASSOCIATIVITY must be 0 and DELOBJ must be 1. If DELOBJ is 3 (default), the original objects are not deleted.

In several situations, the 3DFACE command can be used instead of PLANESURF. This command, fully presented in *Chapter 12, Meshes and surfaces*, creates a planar face only by specifying three or four corners. The advantages with its application are that we can stretch any corner, even with grips, and corners can be anywhere in 3D, independent of the current UCS. For instance, we can create a vertical face easily to simulate a door or window. These faces can be converted to surfaces with the PLANESURF command.

Conversions between 3D objects

Here are two commands to convert objects to solids or surfaces, concluding the creation processes. These commands can be found on the Menu bar, **Modify | 3D Operation**, and on the ribbon, **Home | Solid Editing**.

Converting objects to solids

Here is the command to convert some type of objects to solids.

The CONVTOSOLID command

The CONVTOSOLID command (no aliases) converts the following objects to solids:

- 2D polylines with uniform width and nonzero thickness
- 2D closed polylines with nonzero thickness
- Circles with nonzero thickness
- Closed surfaces without openings between edges
- Closed 3D meshes without openings between edges

The command just prompts for the selection of objects. The DELOBJ variable (with value 1) indicates the deletion of the original objects.

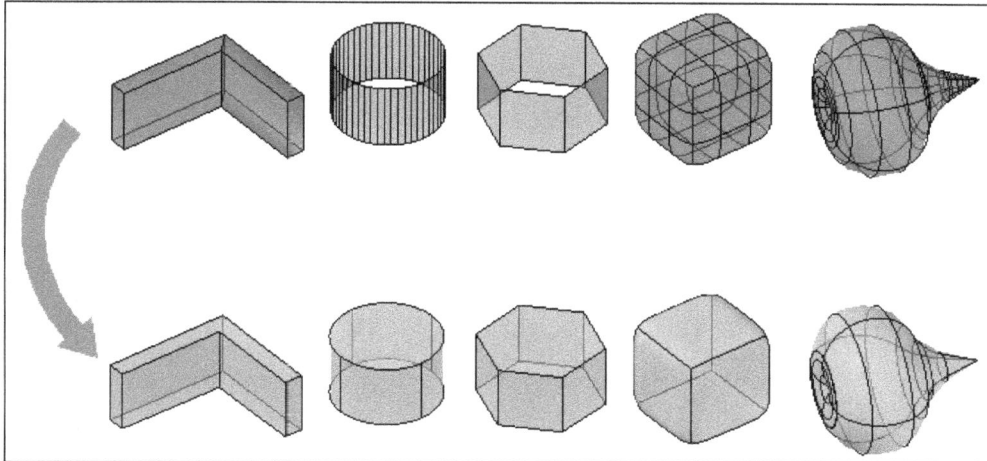

[📝 The converted solids maintain layer, color, and other general properties from the original objects. The current layer is not applied.]

Converting objects to surfaces

The next command converts some type of objects to surfaces.

The CONVTOSURFACE command

The CONVTOSURFACE command (no alias) converts the following objects to surfaces:

- 2D open polylines with nonzero thickness
- Lines and arcs with nonzero thickness
- Regions
- Planar 3D faces
- 2D solid objects (the SOLID command)
- 3D solids
- 3D meshes

The command just prompts for the selection of objects. The DELOBJ variable (with value 1) indicates the deletion of the original objects.

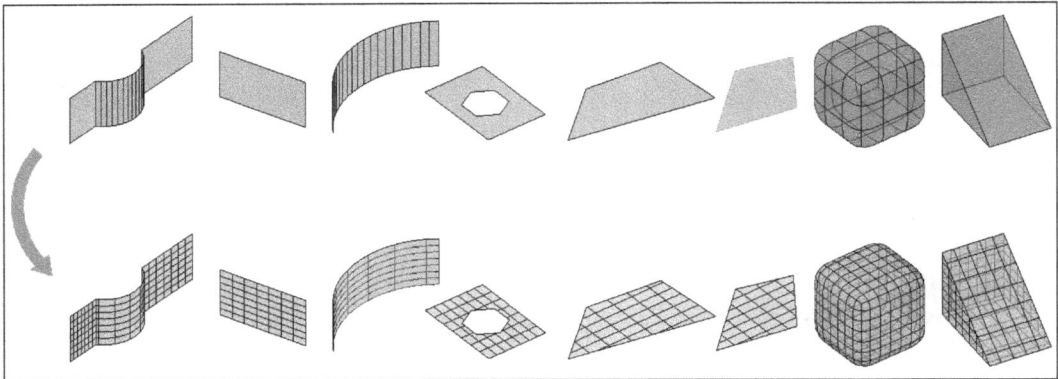

> The converted surfaces maintain layer, color, and other general properties from the original objects. The current layer is not applied.

Summary

This chapter includes all 3D basic shapes or 3D primitives, allowing for the construction of simple 3D models. We presented the BOX, WEDGE, CYLINDER, CONE, SPHERE, TORUS, PYRAMID, and POLYSOLID commands. Also included is PLANESURF, a command for creating planar surfaces, and two commands for converting existing geometry to solids, CONVTOSOLID, or surfaces, CONVTOSURFACE.

In the next chapter, we will see all the general 3D editing commands. The editing commands specific to solids and surfaces are presented in *Chapter 7, Editing Solids and Surfaces*.

6
Editing in 3D

In the first chapter we analyzed the application of AutoCAD editing commands commonly used in 2D, such as MOVE, COPY, SCALE, or ERASE, which work exactly the same way in 3D as in 2D. Others, such as FILLET, CHAMFER, or OFFSET depend on the object plane. Some more, such as MIRROR or ROTATE depend on the current user coordinate system.

In this chapter we present some commands that are specific to 3D operations that can be applied to any object. The editing commands specific to solids and surfaces are the subject of the next chapter.

The topics covered in this chapter are:

- How to rotate objects around an axis
- How to get mirrored objects
- How to align objects
- How to apply multiple, equally spaced copies
- How to modify objects with grips
- How to modify fun faces, edges, and vertices

3D editing commands

The commands that are specific for editing in 3D are rotating around an axis, applying a mirror in relation to a plane, creating equally spaced copies in the three directions, and aligning. The chapter also includes some commands that are less used with objects but that can be useful for editing parts of meshes (explained in *Chapter 12, Meshes and Surfaces*).

The access to some of these commands, besides digitizing their names or alias, can be done on the ribbon, **Home** tab/**Modeling** panel, on the **Modeling** toolbar, and on the **Modify/3D Operations** menu bar.

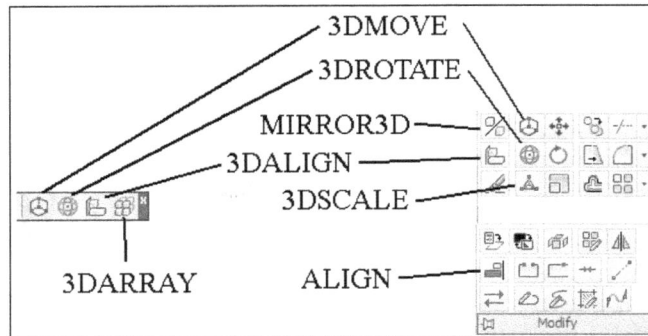

In 3D, Autodesk recommends applying the 3DMOVE, 3DROTATE, 3DALIGN, and 3DSCALE commands available since Version 2007, the reason why they have icons. Unless we are dealing with parts of meshes (the technical term is subobjects), it's faster and easier to apply the MOVE, ROTATE3D, ALIGN, and SCALE commands respectively. All commands except MOVE and SCALE are presented in this chapter.

Rotating in 3D

The ROTATE command allows you to rotate objects in an angle around a base point so that it always refers to the XY working plane of the current user coordinate system. In 3D we are frequently rotating around an axis. There are two commands for this task.

The ROTATE3D command

The ROTATE3D command (neither has an alias nor can it be accessed by any icon or menu bar) allows to rotate the selected objects around an axis defined by two points (default), an angle. It is similar to the ROTATE command except that it defines an axis instead of a point.

The command starts by displaying the current values of the ANGDIR and ANGBASE variables and prompts for the object's selection. The ANGDIR variable states the positive direction for measuring angles, such as counterclockwise or clockwise. The ANGBASE variable indicates the 0 base angle for defining absolute angles.

```
Command: ROTATE3D
Current positive angle:  ANGDIR=counterclockwise  ANGBASE=0
Select objects: Selection
```

By default, we specify the axis of rotation with the help of two points:

```
Specify first point on axis or define axis by
[Object/Last/View/Xaxis/Yaxis/Zaxis/2points]: Point
Specify second point on axis: Point
```

The next step is to specify the relative angle, which is in the positive direction defined by the ANGDIR variable:

```
Specify rotation angle or [Reference]: Angle
```

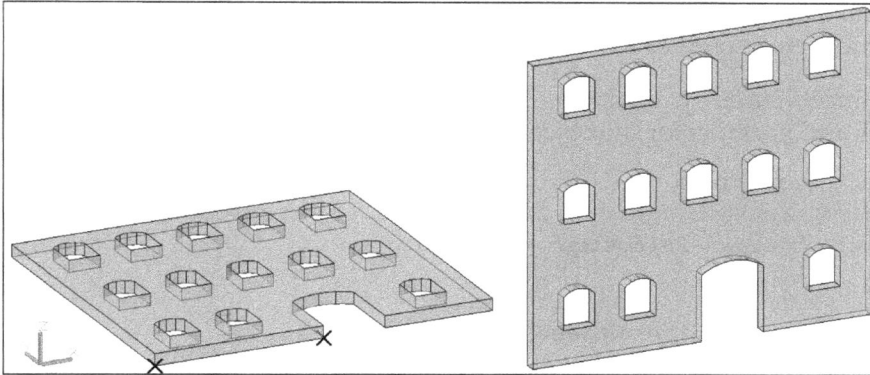

The options for this command are:

- **Object**: The axis of rotation is defined by the selection of a line, an arc, a circle, or an element of a 2D polyline. For selecting a line, the axis is defined by the line position; for arcs and circles, the axis is defined by the normal to its plane that passes through the center; for an element of a polyline, the axis is defined as selecting a line or an arc, depending on the type of element.

- **Last**: This applies the last used axis of rotation.

- **View**: The axis of rotation is defined by the viewing direction (perpendicular to the screen) that passes through a point. It prompts for that point.

- **X axis/Y axis/Z axis**: The axis of rotation is defined by the X axis/Y axis/Z axis direction of the current user's coordinate system that passes through a point. It prompts for that point.

- **Reference**: Instead of specifying an angle's value, this option allows for specifying a current angle and then a new angle. Any of these two angles can be specified by two points.

> The ROTATE3D command is included in the external application GEOM3D.ARX (versions prior to 2013) or GEOM3D.CRX. If this command stops working by giving information to the Unknown command, we must find the respective application and load it with the APPLOAD command.

The 3DROTATE command

The 3DROTATE command (alias 3R) allows a specified value for rotating objects around an axis. This command has been available since Version 2007. It displays a rotate gizmo composed of three circles, which normally define the axis of rotation.

The command starts by displaying the current values of the ANGDIR and ANGBASE variables and prompts for the object's selection:

```
Command: 3DROTATE
Current positive angle in UCS:  ANGDIR=counterclockwise  ANGBASE=0
Select objects: Selection
```

> **Downloading the example code**
>
> You can download the example code files for all Packt books you have purchased from your account at http://www.packtpub.com. If you purchased this book elsewhere, you can visit http://www.packtpub.com/support and register to have the files e-mailed directly to you.

Then, the command places the rotate gizmo at the geometric center of the object or of the selection and prompts for a base point:

```
Specify base point: Point
```

The rotate gizmo is placed at the chosen point and the command asks to pick one of the three axes: red for X, green for Y, and blue for Z:

```
Pick a rotation axis: Selection
```

The selected axis turns yellow and the respective axis is displayed. We may write a value for the angle or specify it with two points. Specifying a first point, the command displays the rotation dynamically:

```
Specify angle start point or type an angle: Value
```

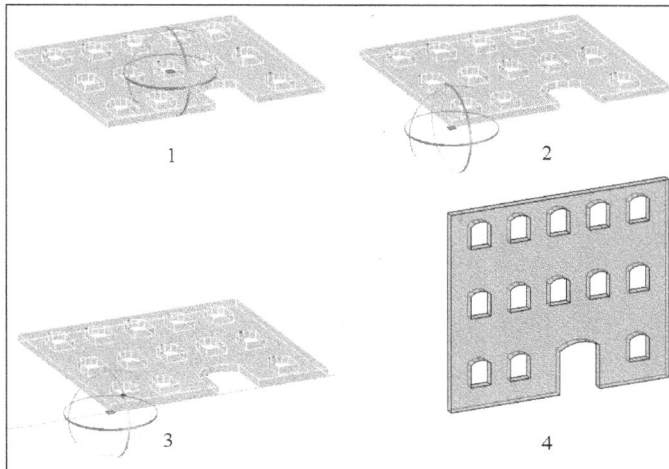

Over the rotate gizmo, we may call the context menu (by right-clicking) and access further options in order to change the gizmo to 3DMOVE and 3DSCALE, for constraining around the X, Y, or Z axis, in order to relocate or align the gizmo and the additional option **Custom Gizmo** for defining a gizmo based on points or an existing object.

> This command, compared with the previous one, doesn't allow for rotating objects around any two points or around an existing object unless we apply the **Custom Gizmo** option from the context menu. And it needs an extra click to perform the rotation.
>
> Related variables to these gizmos, such as the GTAUTO controls with a visual style other than 2D Wireframe in a viewport if the gizmo is displayed upon selection without command, GTDEFAULT controls if 3DMOVE, 3DROTATE, and 3DSCALE commands replace automatically MOVE, ROTATE, and SCALE commands in a viewport with a visual style other than 2D Wireframe, and GTLOCATION controls the initial position for the gizmo.

Exercise 6.1

From a 2D stair, we will create a 3D stair by following these steps:

1. Open the drawing A3D_06_01.DWG. This drawing contains the section of a stair.

2. Create and activate a layer named 3D-STAIR with a different color.

3. By applying PRESSPULL inside the section and specifying 1.2 units for extrusion, the stair is created, but it must be rotated.

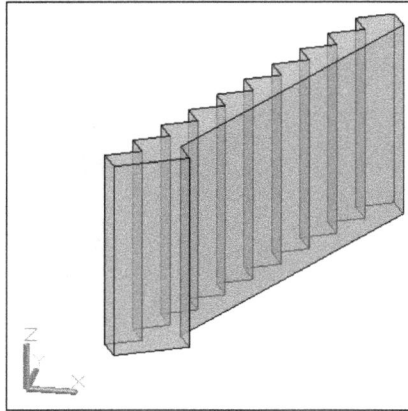

4. To this fallen stair, apply the ROTATE3D command, defining as the axis of rotation the two vertices of the bottom edge, selecting them from the left-hand side to the right-hand side, and specifying an angle of 90 degrees.

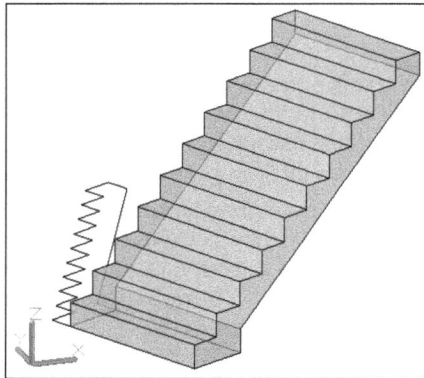

5. Save the drawing with the name A3D_06_01FINAL.DWG.

Mirroring in 3D

AutoCAD uses the next command to mirror objects in relation to a plane. So, instead of using two points to define the symmetry line, as used in the MIRROR command, here we define a symmetry plane.

The MIRROR3D command

The MIRROR3D command (no alias) allows you to mirror objects in relation to a plane defined anywhere in 3D.

As usual, this command starts by requesting the object's selection:

```
Command: MIRROR3D
Select objects: Selection
```

Next, it's up to you to define the symmetry or mirror plane. By default, it asks for three noncollinear points:

```
Specify first point of mirror plane (3 points) or
[Object/Last/Zaxis/View/XY/YZ/ZX/3points] <3points>: Point
Specify second point on mirror plane: Point
Specify third point on mirror plane: Point
```

Finally, we have to decide if we maintain both sets (source and mirrored) or only the mirrored ones:

```
Delete source objects? [Yes/No] <N>: Yes or No
```

The following screenshot shows the source and mirrored sets:

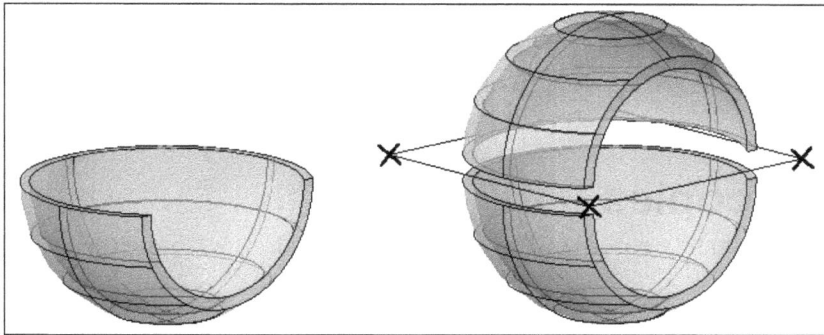

The options for this command are:

- **Object**: The mirror plane is defined by the plane of a planar object to be selected

- **Last**: This option applies the same mirror plane used in the previous MIRROR3D command

- **Z axis**: The mirror plane is defined by a point and its Z direction; it is useful when we want a plane to be perpendicular to some direction

- **View**: This option defines a mirror plane perpendicular to the viewing direction through a point to be prompted

- **XY/YZ/ZX**: The mirror plane is defined by a plane parallel to the XY, YZ, or ZX axis of the current user coordinate system that passes through a point to be prompted

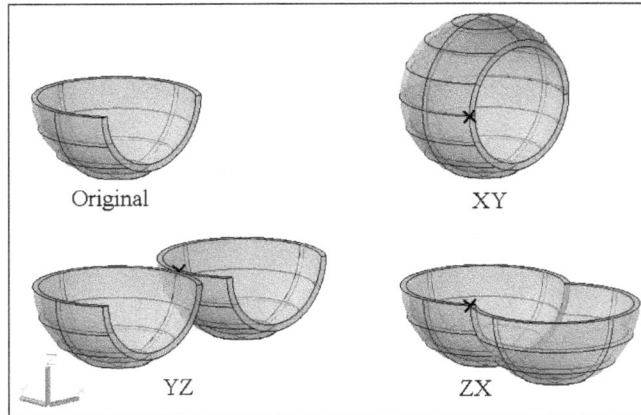

> This command is also included in the external application GEOM3D. ARX (versions prior to 2013) or GEOM3D.CRX. If this command stops working by giving information to the Unknown command, we must find the respective application and load it with the APPLOAD command.

Multiple copies

Generalizing the ARRAY command to the third dimension, we have the 3DARRAY command.

The 3DARRAY command

The 3DARRAY command (alias 3A) is used to create equally spaced multiple copies of the selected objects in a rectangular or polar pattern, by following these steps:

1. We will start by demonstrating the rectangular pattern. After selecting the objects, choose **Rectangular**:

   ```
   Command: 3DARRAY
   Select objects: Selection
   Enter the type of array [Rectangular/Polar] <R>: Rectangular
   ```

2. Define the number of rows along the Y direction of the current user's coordinate system:

```
Enter the number of rows (---) <1>: Number
```

3. Define the number of columns along the X direction:

```
Enter the number of columns (|||) <1>: Number
```

4. Define the number of levels along the Z direction:

```
Enter the number of levels (...) <1>: Number
```

5. Next, specify the distances between equivalent points (for instance, the lower-left corner to the next lower-left corner), starting from the distance between the rows along Y. A negative value places the copied objects along the –Y direction:

```
Specify the distance between rows (---): Value
```

6. Specify the distance between the columns along X:

```
Specify the distance between columns (|||): Value
```

7. Again, specify the distance between columns along X:

```
Specify the distance between levels (...): Value
```

8. If we want to place multiple copies equally spaced around an axis, after selecting objects, choose **Polar**:

```
Command: 3DARRAY
Select objects: Selection
Enter the type of array [Rectangular/Polar] <R>: Polar
```

9. Specify how many sets of the selected objects we want, including the original set:

```
Enter the number of items in the array: Number
```

10. The angle to fill is done next. A positive angle value creates the copies counter clockwise and a negative value creates the same but clockwise:

```
Specify the angle to fill (+=ccw, -=cw) <360>: Value
```

11. As the sets are being placed around an axis, choose between rotating them or not. By default, the sets are rotated:

```
Rotate arrayed objects? [Yes/No] <Y>: Yes or No
```

12. Finally, specify the axis of the polar array by two points:

```
Specify center point of array: Point
Specify second point on axis of rotation:  Point
```

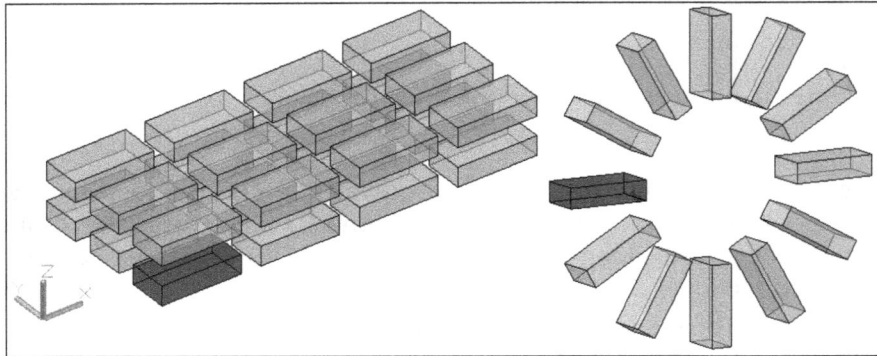

> The changed ARRAY command, since Version 2012, also allows you to create 3D arrays in a rectangular or polar pattern. But, to get the same results as this command, it needs more clicks and options. The advantage of using the ARRAY command is that the array stays associative, that is, we may change the set or the array parameters.
>
> This command is also included in the external application GEOM3D.ARX (versions prior to 2013) or GEOM3D.CRX. If this command stops working by giving information to the Unknown command, we must find the respective application and load it with the APPLOAD command.

Exercise 6.2

The previous two commands will be used to create a roof by following these steps:

1. Open the drawing A3D_06_02.DWG. This drawing contains two beams, one with a rectangular section and the other with a circular section. From these we are going to create the structure of a roof.

2. These elements were created on the plane. First, rotate them so that the rectangular beam has its lower section on the plane. Apply ROTATE3D, and as the axis of rotation, specify the lower vertices of the section and the angle is 90 degrees. Notice that the lower face of the rectangular beam must be on the XY plane.

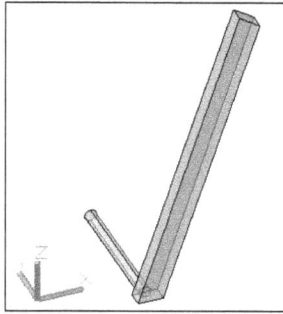

3. To complete the first beam structure, apply the MIRROR3D command three times. For the first application, select the beam, define the mirror plane by the three top vertices, and do not delete the source object. Then, apply the command two more times, selecting one or two beam elements and defining the mirror plane by three endpoints or three midpoints.

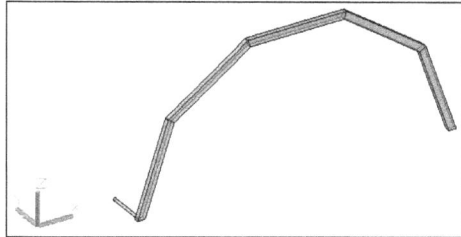

4. To create the first set of circular beams, apply the 3DARRAY command. After selecting the circular beam, set the Polar array. Then specify six items, one of which is an angle of 180 degrees, and consider yes as an answer for the rotation of the arrayed objects (well, as this beam is circular, yes or no are both valid answers). The endpoint of the first axis is obtained by tracking the object from any lowest vertex to any midpoint of the horizontal beam. The endpoint of the second axis is any point along the positive Y axis.

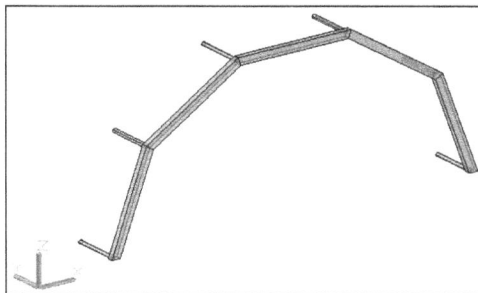

5. Now, apply to all objects the 3DARRAY command, a rectangle, ten rows, one column, and one level; the distance between rows is 2.9.

> As we just want a multiple copy in one direction, instead of applying this command, we could get the same result with Version 2012 or later, the COPY command, the **Array** option, and the ARRAY, -ARRAY, or ARRAYCLASSIC commands. With versions prior to 2012, the ARRAY command would be easier.

6. Finally, delete the last set of circular beams.

7. Save the drawing with the name A3D_06_02FINAL.DWG.

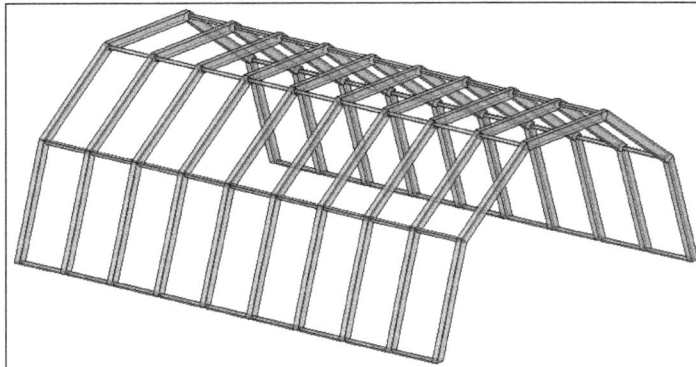

Aligning objects

When we need to align two objects, plane to plane, we may need to provide one move and two rotations. There are two commands that can perform that task easily.

The ALIGN command

The ALIGN command (alias AL) allows you to align two objects by defining a correspondence between three pairs of points. The three source points define a plane on the objects to be aligned. The three destination points define a plane on the target object. After selecting the objects to align (only those that are to be moved and rotated), we specify the first reference point on the objects to be placed:

```
Command: ALIGN
Select objects: Selection
Specify first source point: Point
```

Then, we specify a point cn the target object. These two points define the translation:

```
Specify first destination point: Point
```

We specify the first point of the second pair on the objects to be placed. If, instead of specifying a point we press *Enter*, the command is equivalent to MOVE:

```
Specify second source point: Point
```

We specify the second point of the second pair on the target object. These two points define the first rotation:

```
Specify second destination point: Point
```

We specify the first point of the third pair on the objects to be placed. If, instead of specifying a point we press *Enter*, the command allows you to align only in 2D (one translation and one rotation) and prompts if we want to scale the selected objects based on the two pairs of points:

```
Specify third source point or <continue>: Point
```

We specify the second point of the third pair on the target object. The selected objects are moved and rotated, so both planes are aligned:

```
Specify third destination point: Point
```

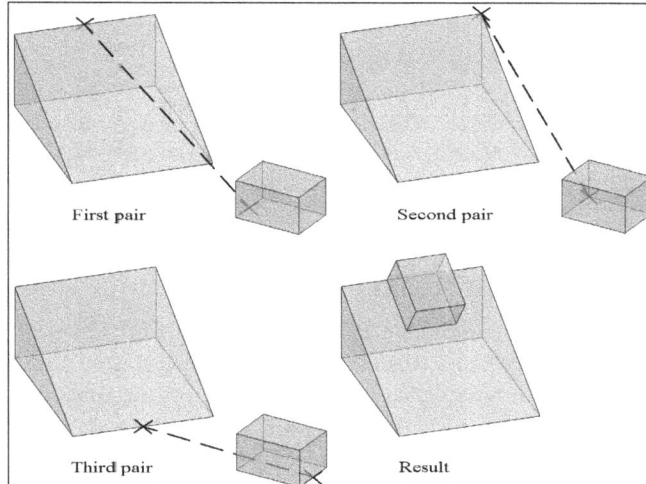

This command can also be quite useful in 2D. With just the definition of two pairs of points, this can be equivalent to one move having one rotation with reference, and eventually one scale with reference.

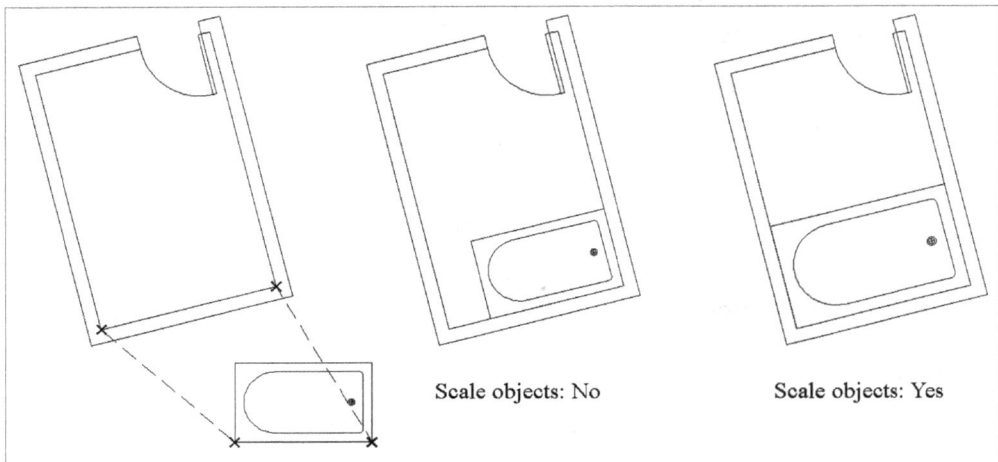

Scale objects: No   Scale objects: Yes

> During the ALIGN command, it is very useful to orbit and zoom in order to specify points with precision.
>
> If the selected objects go inside the target object instead of being outside, it means that the second or the third destination point was specified in the opposite direction.
>
> This command is also included in the external application GEOM3D. ARX (versions prior to 2013) or GEOM3D.CRX. If this command stops working by giving information to the Unknown command, we must find the respective application and load it with the APPLOAD command.

The 3DALIGN command

The 3DALIGN command (alias 3AL) is another command that allows us to align selected objects. This command is different from the ALIGN command in 2 ways: first, we select the three source points, and after the three destination points, the command becomes dynamic, that is, the selected objects follow the cursor as we are specifying the destination points. Second, after selecting the objects to align, we specify the three source points that define the first plane:

```
Command: 3DALIGN
Select objects: Selection
  Specify source plane and orientation ...
Specify base point or [Copy]: Point
```

```
Specify second point or [Continue] <C>: Point
Specify third point or [Continue] <C>: Point
```

Then, we specify the three destination points in the same order, which define the destination plane. The selected objects follow the cursor:

```
Specify destination plane and orientation ...
Specify first destination point:  Point
Specify second destination point or [eXit] <X>: Point
Specify third destination point or [eXit] <X>: Point
```

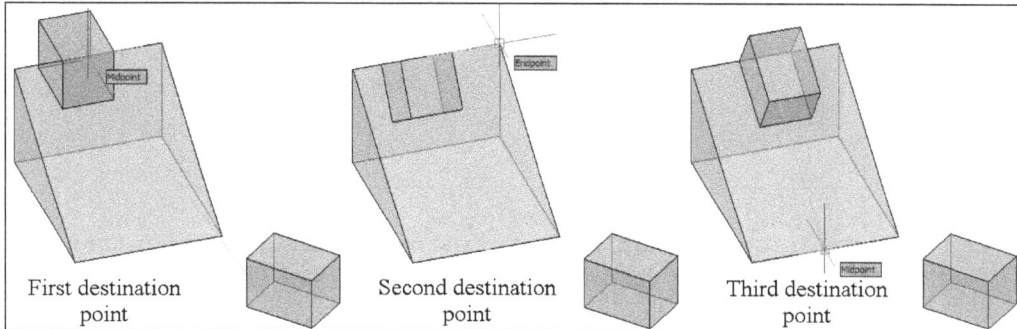

| First destination point | Second destination point | Third destination point |

> If the selected objects are not simple, the fact that they follow the cursor can impair efficiency. While moving the cursor with objects associated to this movement, it is much more difficult to mark points with precision.

Exercise 6.3

We have a part with a rotated face and we want to align to that face some other parts that are to be obtained by extrusion, by following these steps:

1. Open the drawing A3D_06_03.DWG. This drawing consists of a 3D part and some linear objects.

2. First, apply the PRESSPULL command to the offset part with value 15.

3. Repeat the `PRESSPULL` command to each inner circle with value `40`.

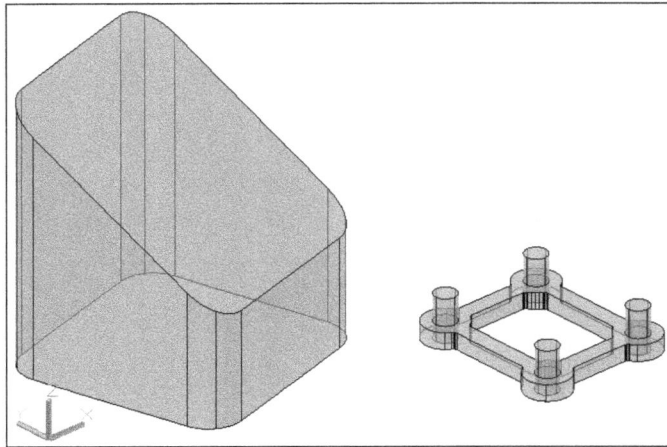

4. Apply the `ALIGN` command and select the five extruded parts. For not to take the chance of selecting the original linear objects, freeze layer `0`.

5. Using object snap tracking and the mid between two points osnap, the first pair of points are the center of the extruded parts to the center of the rotated face.

6. For specifying the second and third pairs of points, choose the midpoints of edges.

7. Save the drawing with the name `A3D_06_03FINAL.DWG`.

Moving objects

The MOVE command works normally in 3D; we can move objects in all directions. Anyway, a command is also available that works preferably with a gizmo.

The 3DMOVE command

The 3DMOVE command (alias 3M) allows you to move objects or parts of objects along the axis or a plane defined by two axes of a specific gizmo. The gizmo represents the three axes and is placed by default at the geometric center of the objects. After selecting the objects to move, the command prompts for the base point and shows the move gizmo:

```
Command: 3DMOVE
Select objects: Selection
```

If we specify a point, the command acts like the MOVE command, prompting for a second point. Normally, we select an axis or a plane:

```
Specify base point or [Displacement] <Displacement>: Axis selection
```

The selected objects are moved along the axis or the plane. We can specify a point or a value along the axis:

```
** MOVE **
Specify move point or [Base point/Copy/Undo/eXit]: Point or Value
```

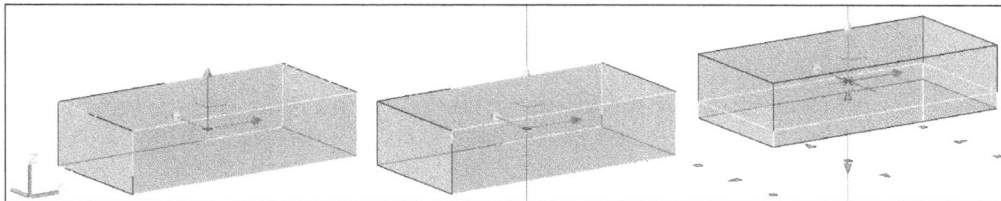

The options for this command are:

- **Base point**: This option prompts for the base point of the move
- **Copy**: This option allows for copying the object along the chosen direction or plane

Compared to the MOVE command, the only advantage of this command is that it allows you to apply some options on the context menu. We can align the gizmo to a user coordinate system to a face or by three points. Then we can move or copy along other directions than the current axes.

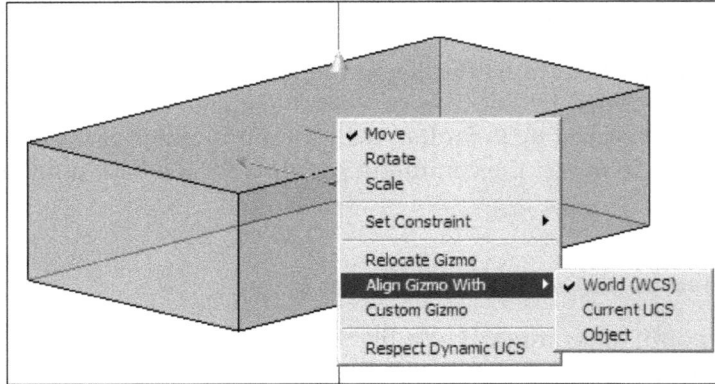

Scaling objects

We have the 3DSCALE command for applying a scale factor to objects or parts of objects. Normally, we apply the SCALE command unless we need to constrain the scale to an axis or a plane for meshes.

The 3DSCALE command

The 3DSCALE command (alias 3S) allows you to apply a scale factor to objects or parts of objects. This command has been available since Version 2010. After selecting the objects, a scale gizmo is displayed:

```
Command: 3DSCALE
Select objects: Selection
```

The scale gizmo is displayed and the command prompts for the base point, for which the scale factor is applied:

```
Specify base point: Point
```

This command prompts for the selection of an axis or plane. For solids and surfaces, we can only indicate the gizmo:

```
Pick a scale axis or plane: Axis selection
```

Finally, we apply a scale factor, either by specifying a point or by digitizing a value:

```
Specify scale factor or [Copy/Reference]: Point or Value
```

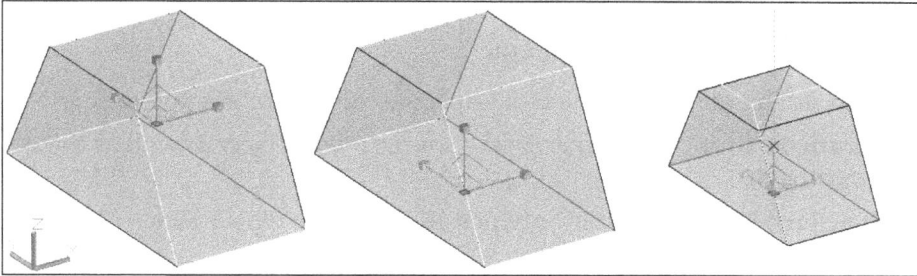

The options for this command are:

- **Copy**: This allows you to copy the object to which the scale is applied
- **Reference**: The scale factor is defined by an existing distance and a new distance, and is similar to the same option of the SCALE command

> For solids and surfaces we can't apply nonuniform scales. Instead, we should apply the SCALE command.

Editing with grips

If we select a solid or a surface without a command, grips are displayed depending on the type of object. Grips are the small blue squares (blue by default) that allow us to perform several editing operations.

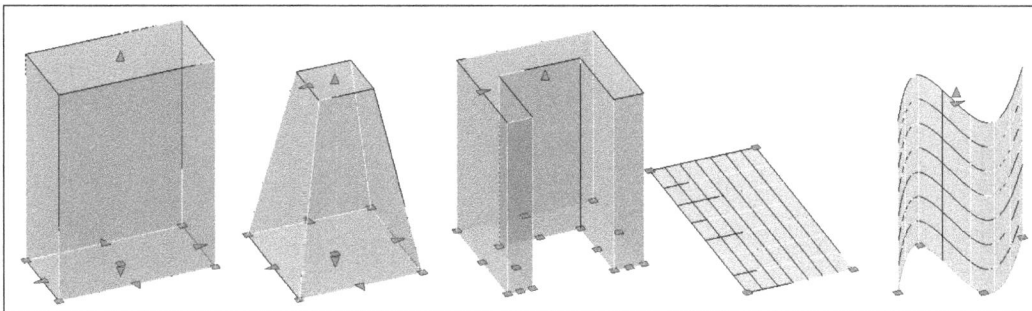

On activating a grip, it turns red (by default) and we can move it elsewhere, thus changing the object. Let's see some examples:

- **Box**: Activating one of the arrow grips, this allows us to adjust the length, width, and height; activating one of the square grips, it allows to change simultaneously length and width.
- **Pyramid**: In addition to the box grips, we can control the top radius.
- **Extruded solid**: We can change the source linear object as well as the height of extrusion.
- **Planesurf**: With grips, it is only possible to move the surface.
- **Surfaces obtained from linear objects**: We can adjust surfaces by moving any of the grips coming from the source linear objects. Additionally, with extruded surfaces it is always possible to adjust the height of extrusion.

Editing subobjects

AutoCAD allows us to modify solids, surfaces, or meshes by moving its subobjects. Subobjects are faces, edges, and vertices.

Subobjects are selected by pressing the *Ctrl* key while selecting a face, edge, or vertex. If we are using an AutoCAD environment based on the ribbon, we may filter the type of subobject at **Home | Selection**.

It is possible to select more than one subobject by keeping the *Ctrl* key pressed.

By using subobjects, it is quite easy to modify the shape of objects. For instance, we may create a four-sided pyramid and then move the two opposite edges to create a rectangular pyramid.

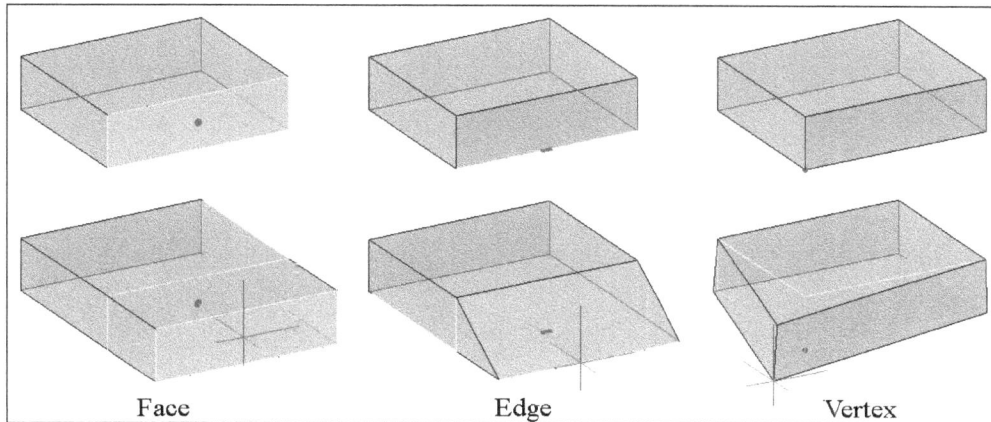

| Face | Edge | Vertex |

Summary

We included in this chapter all the general 3D editing commands that can be applied to several types of objects. We presented the ROTATE3D and 3DROTATE commands to rotate objects around an axis, MIRROR3D to mirror objects in relation to a plane, 3DARRAY to create multiple copies in a rectangular or polar pattern, ALIGN and 3DALIGN to align objects in 3D, and 3DMOVE and 3DSCALE to move and change the scale. We also verified how to edit with grips and subobjects (faces, edges, and vertices).

In the next chapter we will present all commands to edit solids and surfaces.

7
Editing Solids and Surfaces

An important part of any 3D project is combining solids and editing solids and surfaces. Among other useful commands, we can union, subtract, intersect, and cut 3D objects, as well as applying fillets and chamfers to objects edges. In this chapter, we present all the main commands for editing solids and surfaces.

The topics covered in this chapter are as follows:

- Applying unions, subtractions, and intersections
- Cutting and dividing 3D objects
- Filleting and chamfering edges
- Editing faces, edges, and volumes of solids
- Creating copies and offsets from edges
- Projecting objects onto 3D objects
- Adding thickness to surfaces

Composite objects

Until now, we dealt with the creation of 3D objects. Now, let's see how to combine and modify them.

Unions, subtractions, and intersections

Three basic operations to combine objects are unions, subtractions, and intersections. These are called Boolean operations and can be applied to solids, surfaces, and regions.

The commands to perform unions, subtractions, and intersections can be found on the ribbon, **Home | Solid Editing** panel, on the **Modeling** toolbar, and on the Menu bar, **Modify | Solid Editing**:

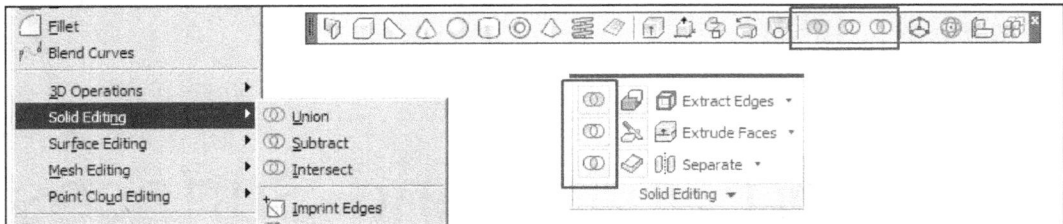

The UNION command

The UNION command (alias UNI) joins two or more objects into one. The command can be applied to solids, surfaces, and regions, but it only joins objects of the same type, that is, it can't join solids with surfaces or regions. It only prompts for the selection of objects to join and returns only one object:

```
Command: UNION
Select objects: Selection
```

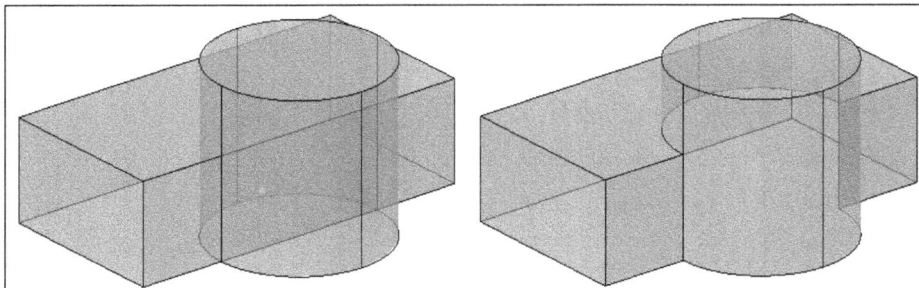

Duplicated volumes or areas are removed.

The command can be applied to solids that do not share a common volume or regions that do not share a common area. In this case, we get one object despite having separated volumes or areas. As it is not possible to join 3D meshes, upon its selection the command prompts for its conversion to solid or surface.

> The application of UNION to surfaces is of minimum interest. In case we need to perform operations on surfaces, the surface commands presented in *Chapter 12, Meshes and surfaces* are more useful.

The SUBTRACT command

The SUBTRACT command (alias SU) allows you to subtract the volume of other objects to a first set of objects. The command can be applied to solids, surfaces, and regions, but doesn't work with objects of different types. The command returns a single object. It starts by prompting the first set of objects:

```
Command: SUBTRACT
Select solids, surfaces, and regions to subtract from ..
Select objects: Selection
```

Then it prompts for the second set of objects. The volume of this set will be removed from the first set:

```
Select solids, surfaces, and regions to subtract ..
Select objects: Selection
```

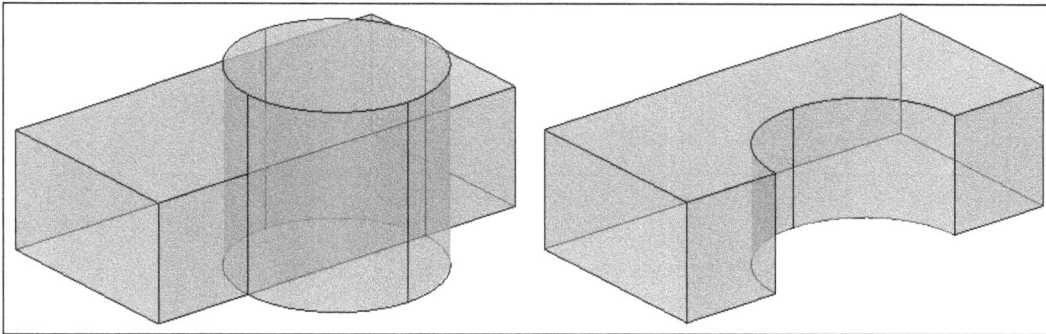

Since Version 2010, this command allows subtracting coplanar surfaces and also subtracting solids to a surface.

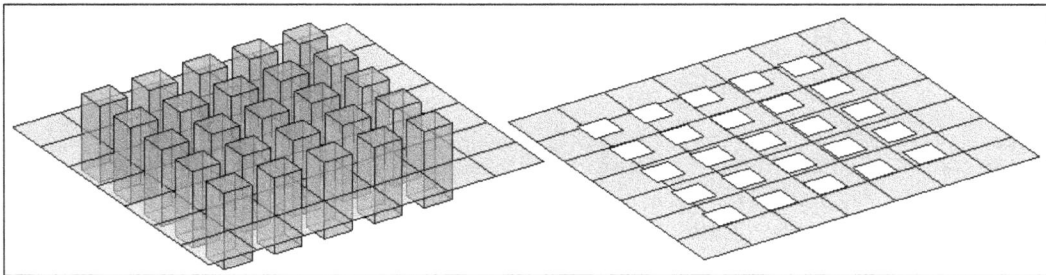

Applying this command to subtract non-coplanar surfaces results in erasing the second set of surfaces and nothing happens to the first set.

The INTERSECT command

The INTERSECT command (alias IN) allows the intersection of objects. Applying to solids returns the common volume, applying to regions and coplanar surfaces returns the common area, and applying to intersecting non planar surfaces returns the intersecting spline. The command just prompts for the selection of objects and returns only one object:

```
Command: INTERSECT
Select objects: Selection
```

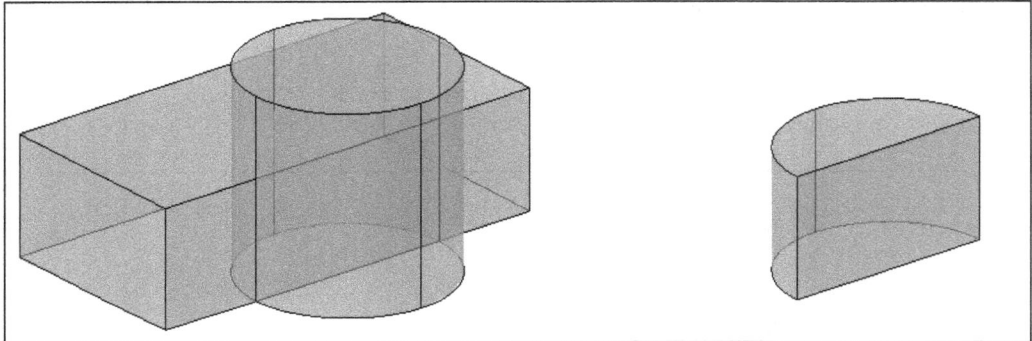

> If there is no common volume, the command returns no object but erases the original objects.
>
> If we want to create a volume corresponding to the intersection of objects but want to maintain the originals, we should apply the INTERFERE command, which will be presented in the next chapter.

Editing composite objects

Under some conditions, we may be able to modify original components of composite objects.

If composite objects, obtained by unions, subtractions, or intersections, have recorded history, that is, kept track of original components, we may edit them by modifying these components.

To record history and have access to original components, we may select the solids to combine and set **History** to **Record** on the **PROPERTIES** palette. By default, all solids don't record history. If we want all solids to record history upon their creation, we have to change the SOLIDHIST variable to 1 (by default, it is 0).

To select a component, we press the *Ctrl* key and click on a surface or edge caused by the component we want to modify. Then we can use grips, the **PROPERTIES** palette, and some editing commands, such as MOVE or 3DMOVE, to edit the component. The composite object updates immediately.

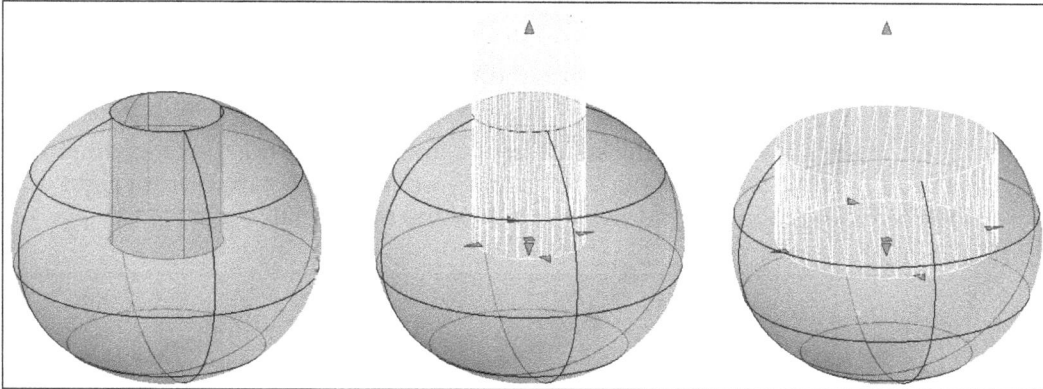

The SHOWHIST variable specifies if component objects are displayed even if we don't select them with the *Ctrl* key. With value 2, component objects are always displayed; with value 1, it depends on individual properties; with value 0, they are never displayed.

> The need for changing components is rare. Normally we don't maintain history records, thus decreasing memory consumption.

The BREP command

The BREP command (no alias) allows removing history records and edition through grips or the **PROPERTIES** palette for the objects to be selected. It only prompts for the selection of composite solids, solid primitives, or solids obtained from linear objects.

All selected solids become generic 3D solids.

Exercise 7.1

We are completing the house initiated in *Chapter 4, Creating Solids and Surfaces from 2D*.

1. Open the drawing A3D_07_01.DWG. This drawing contains the solids of the one-floor house created from a 2D drawing.

2. A good view to select all objects is the **Front** view, accessed by the view cube or another process.

3. First, unite all walls and the solids above doors. Apply the UNION command, selecting all 18 solids on the 3D-WALLS by opening a rectangle from right to left (crossing) above windows and below the upper slab.

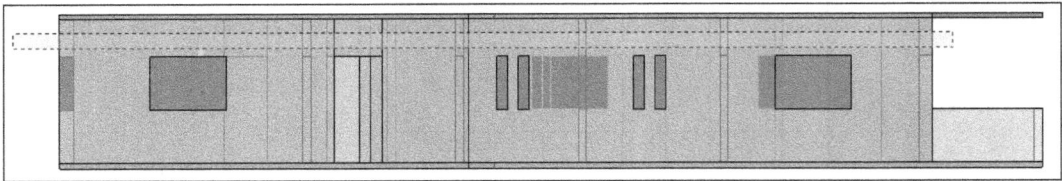

4. We apply SUBTRACT, select the object walls, press *Enter*, select all windows from left to right, and include all the windows completely.

5. Orbiting and eventually hiding the upper slab, we can verify the model correction.

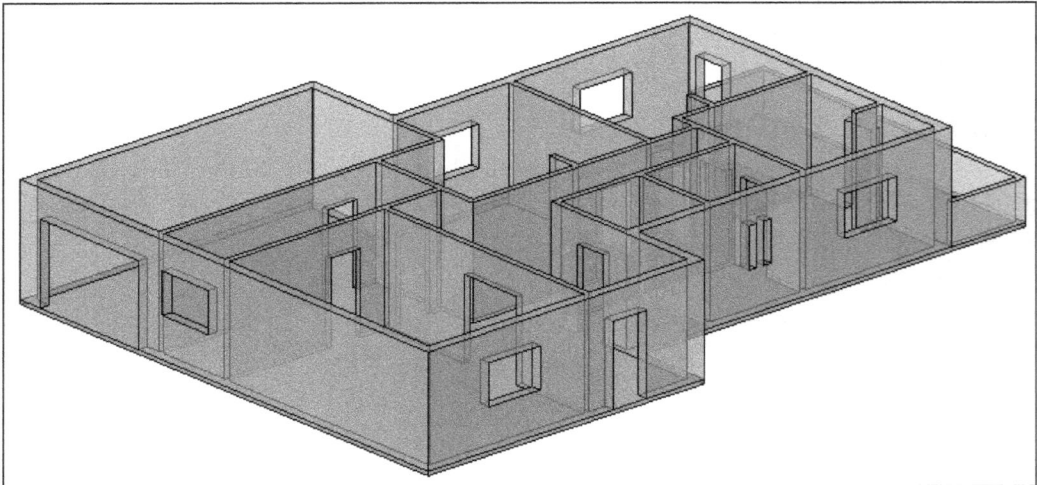

6. Save the drawing with the name A3D_07_01FINAL.DWG.

Exercise 7.2

We are going to continue the cathedral by applying some unions and subtractions.

1. Open the drawing A3D_07_02.DWG. This drawing contains the primitive solids created in *Exercise 5.4* of *Chapter 5, 3D Primitives and Conversions*.

2. Apply the UNION command and select all the boxes, the two pyramids, the rear cylinder and cone, and the torus. All these objects become a single one:

3. To create the buttresses, apply the SUBTRACT command, select all the six larger wedges, press the *Enter* key, and select all six smaller wedges. The result is a single object for all buttresses, but we don't need these as separate objects:

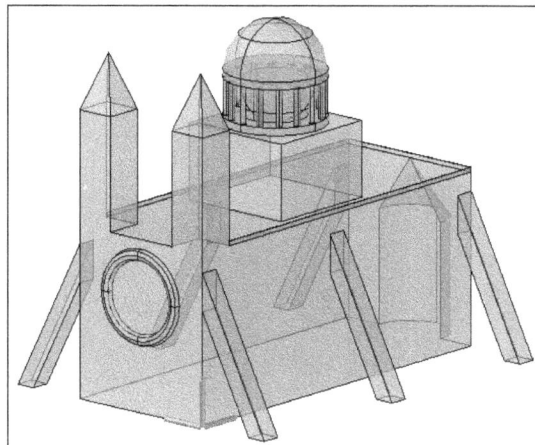

4. We Save the drawing with the name A3D_07_02FINAL.DWG. Later we will cut the summit and make a hollow in the big block.

Cutting objects

Sometimes we need to cut 3D objects. Here is a command to perform this task. The PROJECTGEOMETRY command (presented later in this chapter), although not being its main function, also allows cutting 3D objects.

Cutting solids and surfaces

Here is the most used command to cut solids and surfaces.

The SLICE command

The SLICE command (alias SL) allows cutting solids and surfaces by a plane or an existing surface. We can maintain both parts or select one. We can also access this command on the ribbon, **Home | Solid Editing**, and on the Menu Bar, **Modify | Solid Editing**.

We start the command by selecting the objects to cut:

```
Command: SLICE
Select objects to slice: Selection
```

We have to define the cutting plane. By default, we define a vertical cutting plane, specified by two points:

```
Specify start point of slicing plane or [planar Object/Surface/Zaxis/
View/XY/YZ /ZX/3points] <3points>: Point
Specify second point on plane: Point
```

Then, we specify a point to indicate which part we want to maintain:

```
Specify a point on desired side or [keep Both sides] <Both>: Point
```

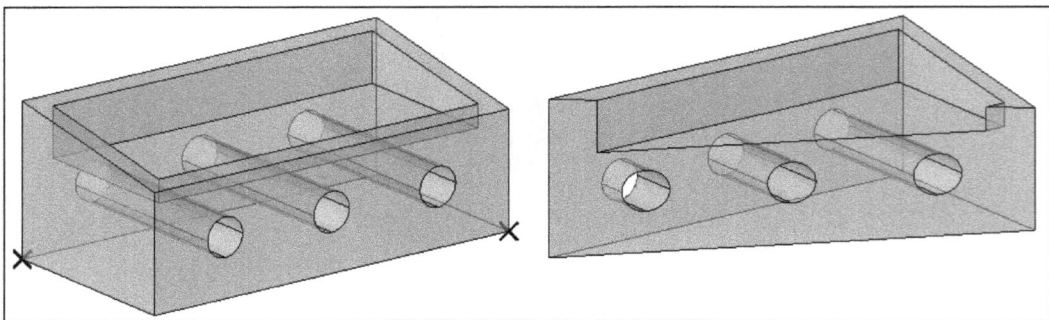

Several useful options are available for this command:

- **planar Object**: The cutting plane is defined by a planar object to be selected, such as a circle, an ellipse, an arc, a 2D spline, or a 2D polyline.

- **Surface**: This is the only option that allows you to specify a non-planar cutting object. The selected surface must intersect the whole object to be sliced.

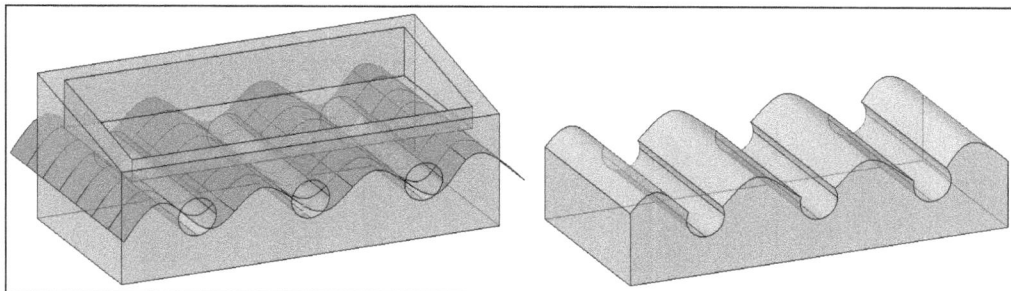

- **Zaxis**: The cutting plane is defined by a point on that plane and a second point that specifies the normal to the plane.

- **View**: The cutting plane is parallel to the current view that passes through a point to be specified.

- **XY/YZ/ZX**: The cutting plane is parallel to the XY, YZ, or ZX plane that passes through a point to be specified.

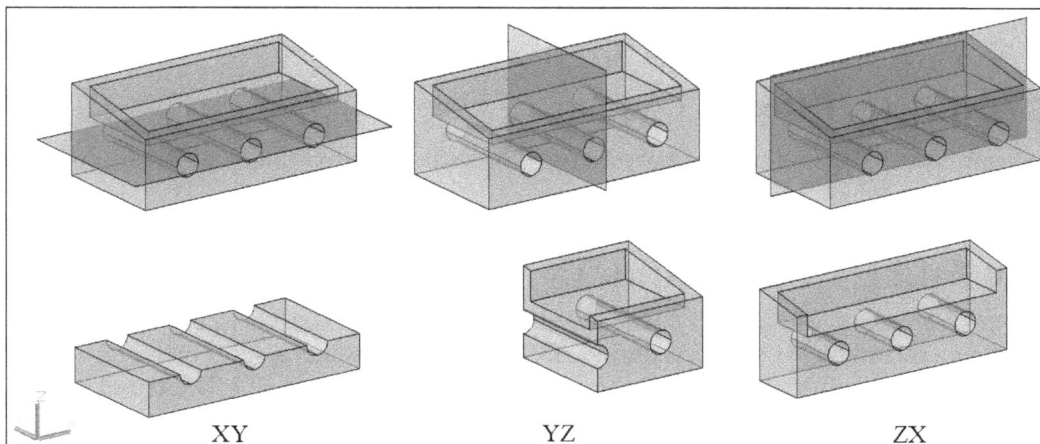

| XY | YZ | ZX |

- **3point**: The cutting plane is defined by three non-linear points to be specified.

- **Both**: When prompting for a point that specifies the part to maintain, this option allows maintaining both parts side-by-side.

Exercise 7.3

We are cutting the summit of the cathedral.

1. Open the drawing A3D_07_03.DWG.

2. Zooming in to the sphere, apply the SLICE command to it, specifying an XY cutting plane that passes through its center. Define a point above the plan to maintain the top hemisphere.

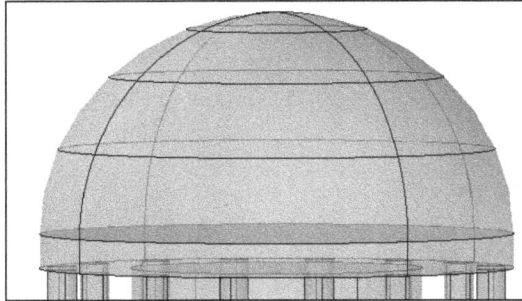

3. Save the drawing with the name A3D_07_03FINAL.DWG. Later we will finish this building.

Filleting and chamfering

Two frequent operations when modeling in 3D are rounding and chamfering edges.

Applying fillets

To round or to fillet edges, we have two commands. The first one is well known from 2D.

The FILLET command

The FILLET command (alias F), besides filleting two linear objects in 2D, allows you to fillet or round edges of solids and surfaces. When we select a solid or surface edge, the command changes the usual prompts and adapts to fillet in 3D. The command starts by displaying the information about the current settings, only important for 2D, and prompts for the selection of the first object:

```
Command: FILLET
Current settings: Mode = TRIM, Radius = 0.0000
Select first object or [Undo/Polyline/Radius/Trim/Multiple]: Edge
selection
```

When we select a solid or surface edge, the command doesn't prompt for the second object, instead it prompts for the radius:

```
Enter fillet radius or [Expression]: Value
```

After that, we click on all edges to be filleted, one-by-one, but only from the selected solid or surface. The edge picked for the object selection already belongs to the selection. After its selection, all edges are filleted and the command displays that information:

```
Select an edge or [Chain/Loop/Radius]: Edge selection
4 edge(s) selected for fillet.
```

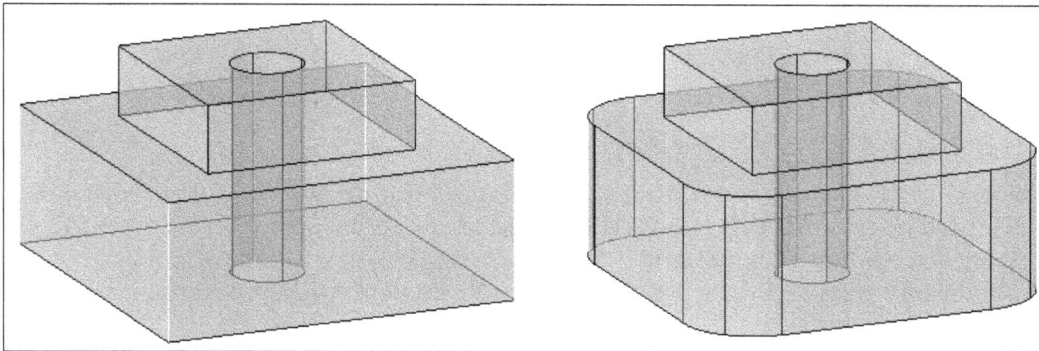

> If the edge is internal, fillet adds material; if external, fillet removes material.

Options for 3D:

- **Chain**: With this option, we can click on an edge and all tangential edges are also selected. There cannot exist any corners.

- **Edge**: This option replaces **Chain** and **Loop** if one of these modes is activated and allows us to turn back to individual selection.

- **Loop**: This option allows for a sequence of edges even not tangential. As an edge is shared by two faces, it prompts to accept the highlighted face or choose next.

- **Radius**: We can change the radius to apply to the next selected edges. It is possible to apply a different radius with only one command:

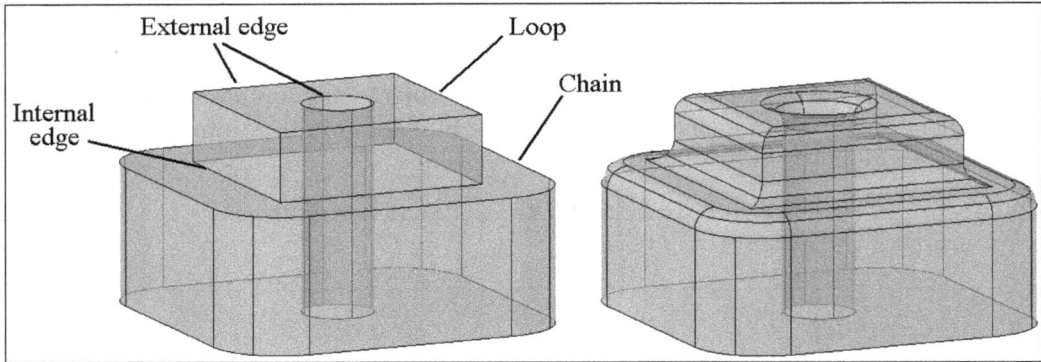

The FILLETEDGE command

The FILLETEDGE command (no alias) is available since Version 2011 and represents a dynamic way for filleting edges. There is an icon for this command on the **Solid Editing** toolbar and ribbon, **Solid | Solid Editing**. This command cannot be applied to surfaces and does not allow the application of a different radius within the same command.

The command displays the current default radius and prompts for an edge selection, which is repeated until *Enter* is pressed. A preview is applied to the model:

```
Command: FILLETEDGE
Radius = Default
Select an edge or [Chain/Loop/Radius]: Selection
Select an edge or [Chain/Loop/Radius]: Selection
2 edge(s) selected for fillet.
```

The solid is displayed transparently. We press *Enter* to accept the default radius, apply the **Radius** option, or click-and-drag the small triangular grip to dynamically specify the radius:

```
Press Enter to accept the fillet or [Radius]: Grip dragging
>>Specify radius or [eXit]:
```

Finally, we press *Enter* to end the command:

```
Press Enter to accept the fillet or [Radius]: Enter
```

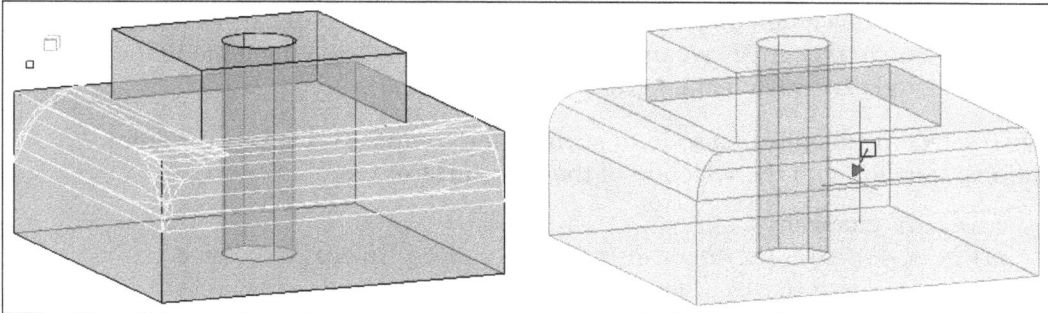

The options for this command are similar to the same options of the previous command:

- **Chain**: With this option we can click on an edge and all tangential edges are also selected. A corner stops this selection.
- **Edge**: This option replaces **Chain** and **Loop** if one of these modes is activated and allows us to turn back to individual selection.
- **Loop**: This option allows for a sequence of edges even not tangential. As an edge is shared by two faces, it prompts to accept the highlighted face or choose next.
- **Radius**: We can change the default radius for all selected edges.

The FILLETRAD3D variable keeps the last used radius for this and the SURFFILLET commands. The PREVIEWCREATIONTRANSPARENCY variable controls the preview transparency.

> The FILLET command has two important advantages: it is possible to apply a different radius within the same application, and avoiding dynamism is faster.

Applying chamfers

Instead of rounded edges, it is also common to find cut edges in engineering or architecture. There are two commands for this task, the first well known from 2D.

The CHAMFER command

The CHAMFER command (alias CHA), besides its application in 2D, allows chamfering edges of solids and surfaces. The effect is the same as applying a diagonal cut. Instead of a radius when filleting, here we must specify two distances that may be different. Due to this, in 3D we must select a base surface to which the first distance is applied, and can only chamfer edges on that surface.

When selecting an edge of a solid or surface, the command displays different options, starting with the selection of the base surface:

```
Command: CHAMFER
(TRIM mode) Current chamfer Dist1 = 0.0000, Dist2 = 0.0000
Select first line or [Undo/Polyline/Distance/Angle/Trim/mEthod/
Multiple]: Selection
Base surface selection...
```

As the selected edge is shared by two surfaces, we have to specify which the base surface is, using the **Next** or **OK** options:

```
Enter surface selection option [Next/OK (current)] <OK>: Next
Enter surface selection option [Next/OK (current)] <OK>: Enter
```

We specify the chamfer distance measured along the base surface:

```
Specify base surface chamfer distance or [Expression]: Value
```

We specify the chamfer distance measured along the other surface, by default, equal to the previous distance:

```
Specify other surface chamfer distance or [Expression] <Default>:
Value
```

We select the edges to chamfer, one-by-one. The edge used for an object's selection is not automatically part of the selection:

```
Select an edge or [Loop]: Selection
```

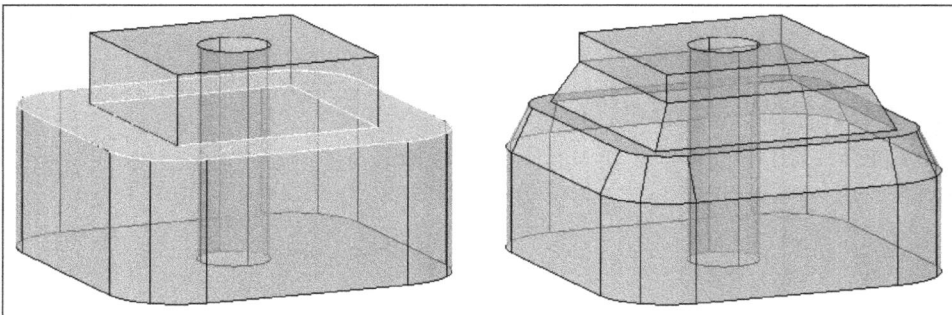

Only two options are available for this command:

- **Loop**: This option allows for selecting a sequence of edges just by selecting one

- **Edge**: This option replaces **Loop** and allows us to turn back to individual selection

> If we want to apply chamfer to other edges not belonging to the base surface, or if we need to specify different distances, we have to end and apply the command again.

The CHAMFEREDGE command

As with fillets, we have a second command to apply chamfers, this one being more dynamic.

The CHAMFEREDGE command (no alias) is available since Version 2011 and represents a dynamic way for chamfering edges. There is an icon for this command on the **Solid Editing** toolbar and ribbon, **Solid | Solid Editing**. This command cannot be applied to surfaces.

The command displays the current default distances and prompts for an edge selection, all on the same face, and is repeated until pressing *Enter*. A preview is applied to the model:

```
Command: CHAMFEREDGE
Distance1 = Default, Distance2 = Default
Select an edge or [Loop/Distance]: Selection
Select another edge on the same face or [Loop/Distance]: Selection
Select another edge on the same face or [Loop/Distance]: Selection
Select another edge on the same face or [Loop/Distance]: Selection
Select another edge on the same face or [Loop/Distance]: Enter
```

The solid is displayed transparently. We press *Enter* to accept the default distances, apply the **Distance** option, or click-and-drag the small triangular grips to dynamically specify the distances:

```
Press Enter to accept the chamfer or [Distance]: Grip dragging
>>Specify distance or [eXit]:
```

Finally, we press *Enter* to end the command:

```
Press Enter to accept the chamfer or [Distance]: Enter
```

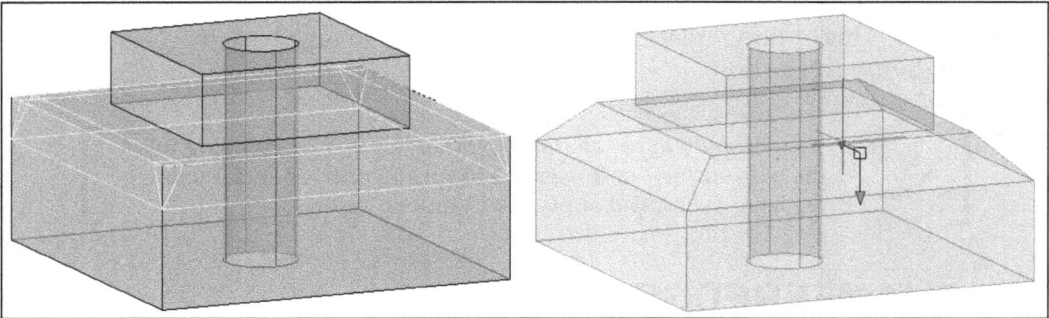

The options for this command are similar to previous commands:

- **Loop**: This option allows for a sequence of edges, even ones not tangential. As an edge is shared by two faces, it prompts us to accept the highlighted face or choose next.
- **Distance**: We can change the default distances.

The PREVIEWCREATIONTRANSPARENCY variable controls the preview transparency.

> The CHAMFER command has two important advantages: it is possible to apply different distances within the same application, and avoiding dynamism is faster.

Exercise 7.4

We are going to fillet and chamfer some edges to the bathtub.

1. Open the drawing A3D_07_04.DWG.
2. Apply the FILLET command with a radius of 0.03 to all edges coming from the smaller box that was subtracted.

3. To the bottom edges, apply the CHAMFER command with distances 0.05 and 0.1. The **Loop** option is an easy way to select all four edges.

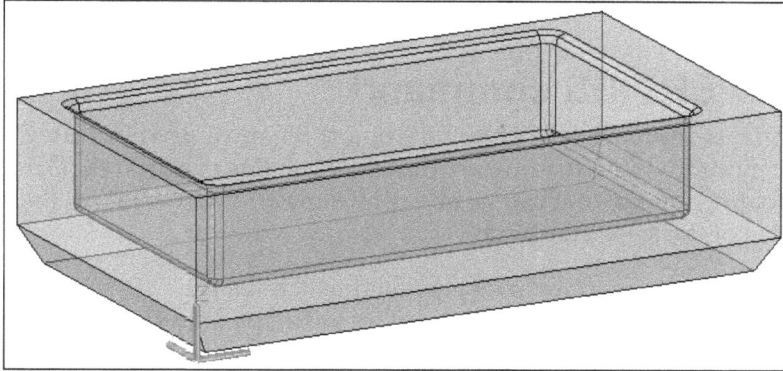

4. Finally, apply the FILLET command to all top exterior edges and vertical edges, and assigning a radius of 0.06.

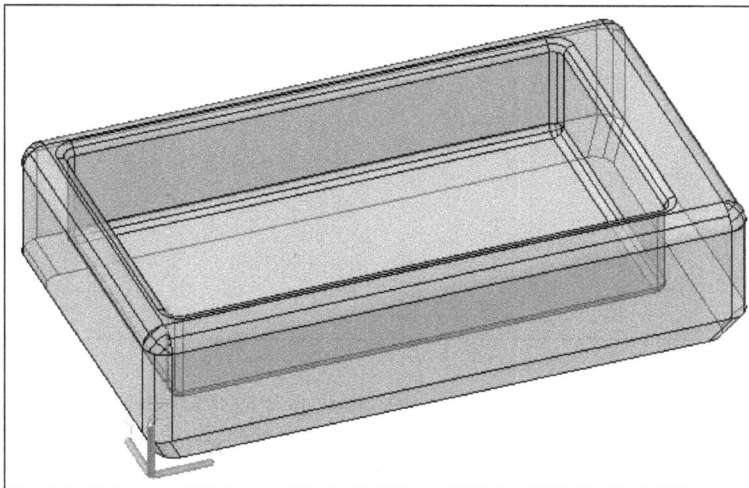

5. Save the drawing with the name A3D_07_04FINAL.DWG.

Editing solids and surfaces

Here are the commands to edit solids and surfaces, namely, changing parameters with the **PROPERTIES** and **QUICKPROPERTIES** palettes and the SOLIDEDIT command with several useful options.

Changing parameters

After a solid or surface creation, it's very easy to change its parameters with the **PROPERTIES** or **QUICKPROPERTIES** palettes.

The PROPERTIES command

The PROPERTIES command (alias *Ctrl + 1*), one of the most used in AutoCAD, is also excellent for modifying the parameters that define solids or surfaces. The command displays a floating palette that can be always visible and that displays the parameters and properties of the selected object.

Selecting a solid, there are three panels related to 3D, namely, **3D Visualization**, **Geometry**, and **Solid History**. The first one includes the material and shadow display only important for renderings and the last one includes history parameters for composite objects. As in 2D, under **Geometry**, we can verify and modify all relevant parameters of the solid or surface selected. By selecting a surface, besides **Visualization** and **Geometry**, two additional panels are available, namely, **Surface Associativity** and **Trims**, where we control associativity between surfaces and original lines and trims applied to surfaces.

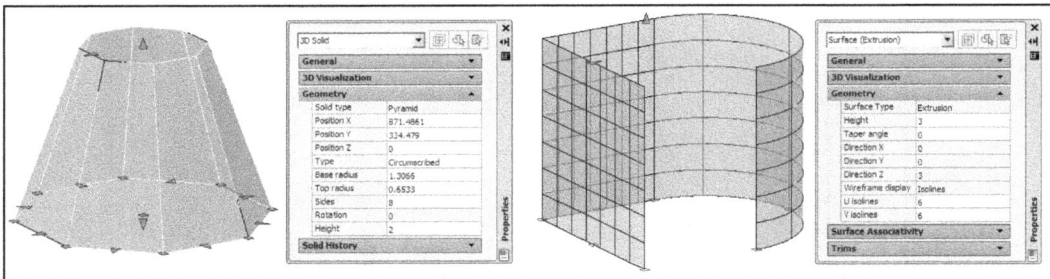

> The easiest method to work with the **PROPERTIES** palette is, if there is no space in the drawing area to maintain its visibility, to apply *Ctrl + 1* to turn it on or off. *Ctrl + 1* actually calls the PROPERTIES or the PROPERTIESCLOSE commands, depending on the palette visibility.

The QUICKPROPERTIES command

The QUICKPROPERTIES command (alias QP) also displays a properties palette, but one that is condensed to the most important parameters and near the object. For most objects, this is the command displayed when double-clicking over that object.

Another way to apply the **Quick Properties** palette is to turn on the **QP | Quick Properties** button on the Status bar (or press *Ctrl + Shift + P*). With this mode activated, it is enough to select an object, without command, and the palette is immediately displayed.

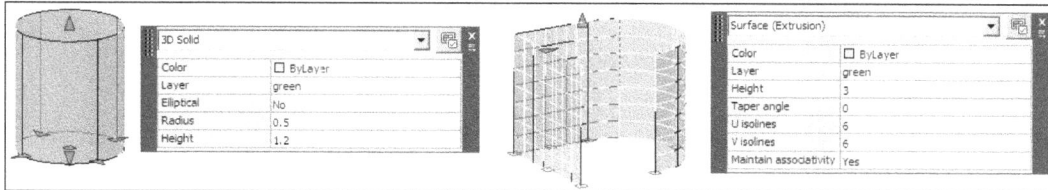

> It is possible to configure which parameters are displayed for each object type, by accessing the CUI command and the **Quick Properties** item. A shortcut is to press the top-right button **Customize** on the palette and we are driven to the proper **Quick Properties** object.

Operations with faces, edges, and volumes of solids

When dealing with solids, it is frequently needed to modify faces, edges, and volumes.

The SOLIDEDIT command

The SOLIDEDIT command (no alias) allows us to modify faces, edges, and volumes (here called bodies) of solids. To apply this command, it is faster to use the **Solid Editing** toolbar or the ribbon, **Home | Solid Editing**; we jump directly to the wanted option.

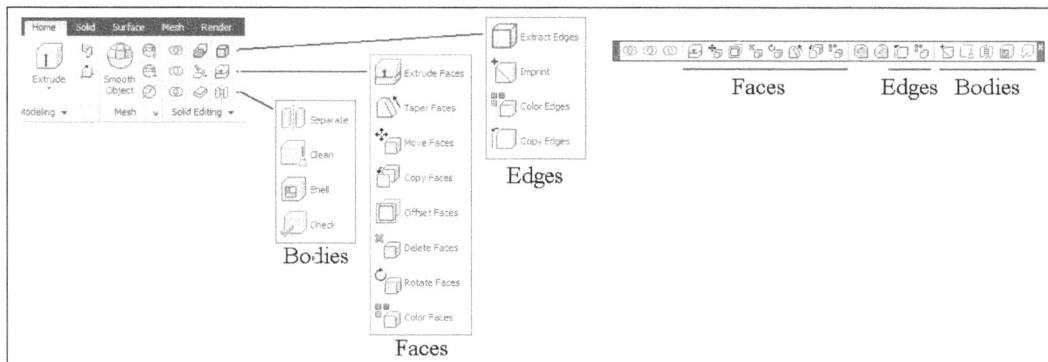

The interaction with the command starts by displaying whether the verification of solids accuracy is turned on (the SOLIDCHECK variable):

```
Command: SOLIDEDIT
Solids editing automatic checking: SOLIDCHECK=1
```

Next, the main command options are displayed. **Undo** undoes the last command operation and **eXit** ends the command. The remaining options are presented next:

```
Enter a solids editing option [Face/Edge/Body/Undo/eXit] <eXit>:
Option
```

The Face option

Upon entering the **Face** option, we have several additional options. **Undo** undoes the last applied option and **eXit** returns to the previous level. When selecting faces, **Undo** and **Remove** options are available, allowing us to undo the last selection operation or entering the mode to remove faces from the selection:

```
Enter a face editing option
[Extrude/Move/Rotate/Offset/Taper/Delete/Copy/coLor/mAterial/Undo/
eXit] <eXit>: Option
```

> We must be careful when selecting faces. If we click an edge, both faces sharing that edge are selected.

The options are as follows:

* **Extrude**: This option allows you to extrude faces. It prompts for faces selection, the height of extrusion, and the angle of taper for extrusion. A positive value for height adds volume, while a negative value removes volume. An angle not equal to zero allows us to taper the new volume.

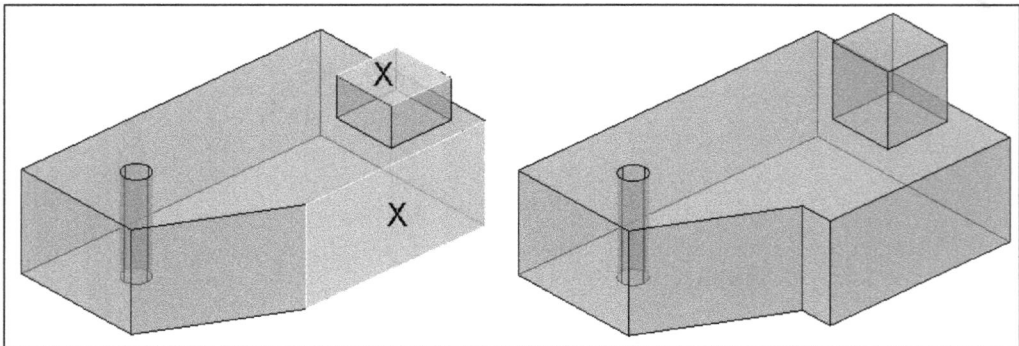

- **Move**: This option allows moving of faces. After selection, it prompts for displacement, for instance, with two points similar to the MOVE command.

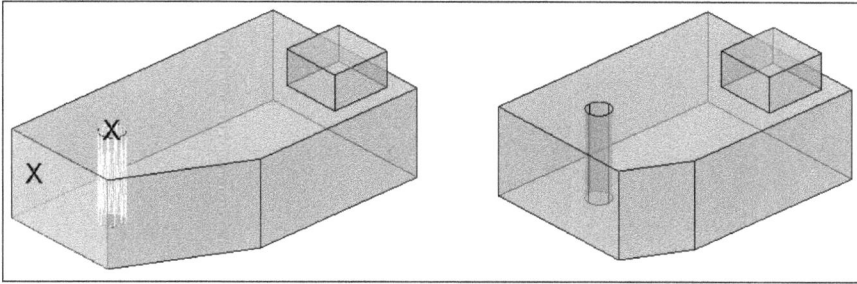

- **Rotate**: This option allows the rotation of faces. After selection, it prompts for the axis of rotation and the angle of rotation. The axis of rotation, by default, is specified by two points, with additional options allowing us to use an existing object or define an axis parallel to the X, Y, or Z axis of the current UCS.

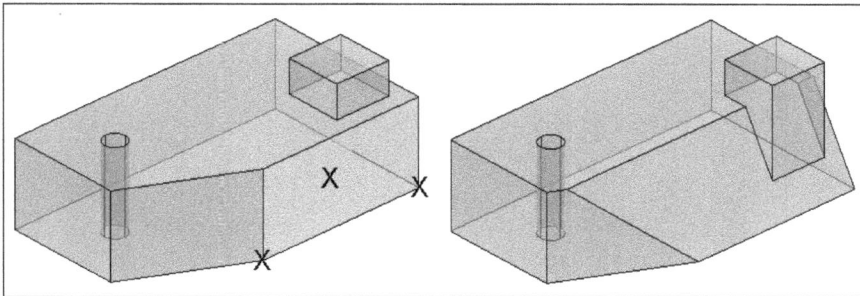

- **Offset**: This option allows the offsetting of faces or the creation of parallel faces. After selection, it prompts for the offset distance. A positive value adds volume, while a negative value removes volume. If selecting continuous faces (for instance, on the common edge), faces are extended so that the corner is corrected.

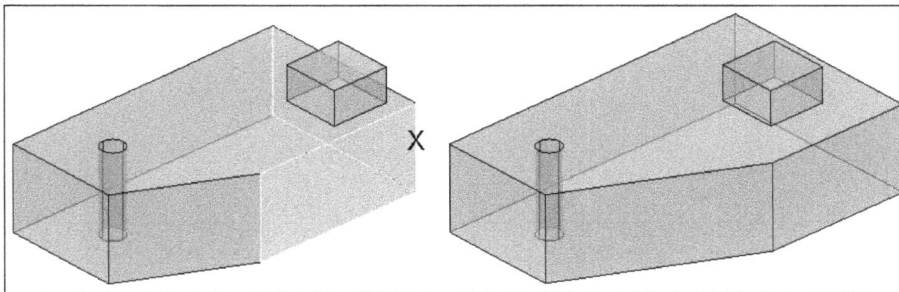

- **Taper**: This option allows us to taper faces. After selection, it prompts for two points that define the taper axis and the taper angle. With positive values, end section is reduced, while with negative values, end section is increased:

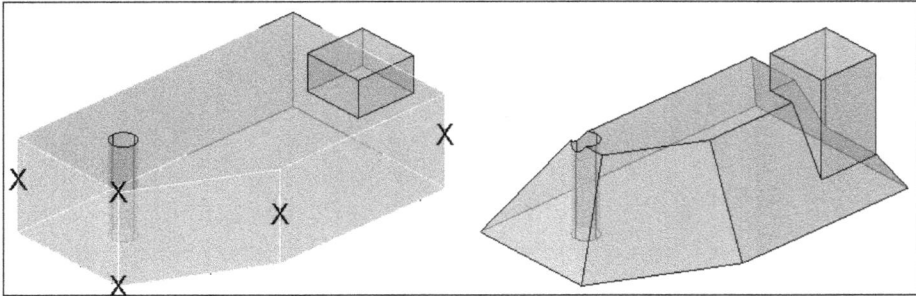

- **Delete**: This option allows the deletion of faces. If AutoCAD knows how to fill the space, faces will be deleted.

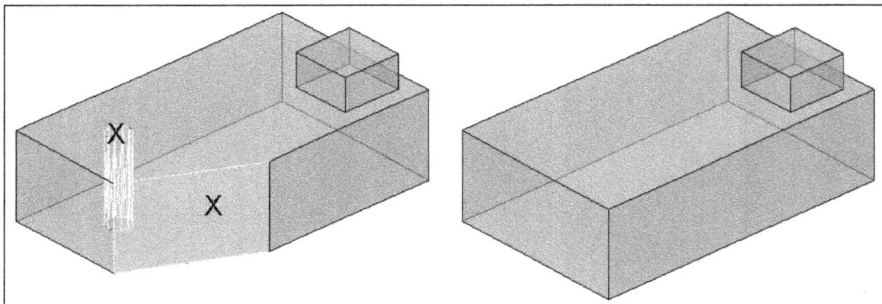

- **Copy**: This option allows the copying of faces. Regions are created for all planar faces, else surfaces are created for all curved faces. After selection, it prompts for two points that define the displacement.

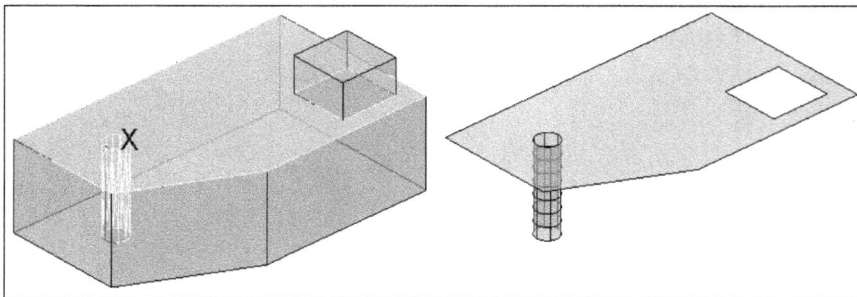

- **coLor**: This option allows changing the color of selected faces. After selection, it shows the **Select Color** dialog box to choose the color to apply.

- **mAterial**: This option allows the application of a material to selected faces, so in render, we can have different materials applied to a single object. After selection, it prompts for the name of the material to apply. This option has no icon.

> The **coLor** and **mAterial** options have limited utility. To apply materials to faces, there are other possibilities for which we don't need to type the material name.

The Edge option

When choosing the **Edge** option, besides the **Undo** and **eXit** options, we have two more options. When selecting edges, **Undo** and **Remove** options are available, allowing us to undo the last selection operation or entering the mode to remove edges from the selection:

```
Enter an edge editing option [Copy/coLor/Undo/eXit] <eXit>: Option
```

- **Copy**: This option allows the copying of edges. Depending on selected edges, arcs, circles, ellipses, lines, or splines are created. After selection, it prompts for two points that define the displacement.

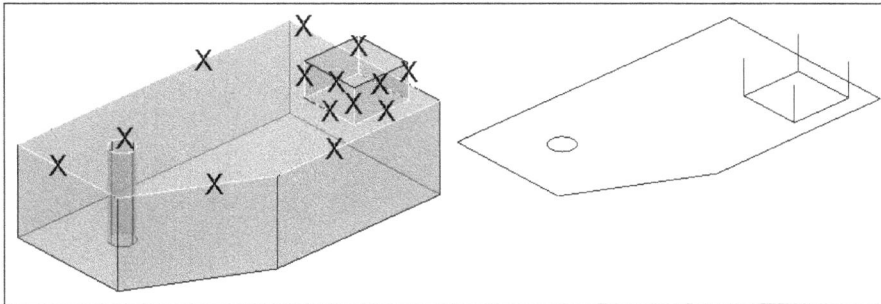

- **coLor**: This option allows us to change the color of selected edges. After selection, it shows the **Select Color** dialog box to choose the color to apply.

The Body option

The body represents the volume of a solid. Besides the **Undo** and **eXit** options, we have five more options:

```
Enter a body editing option
[Imprint/seParate solids/Shell/cLean/Check/Undo/eXit] <eXit>: Option
```

- **Imprint**: This option allows you to create edges from intersections of the selected solid with linear entities, regions, or solids. After one solid selection, it prompts for an object to imprint and then this object is erased. These two prompts are repeated until the *Enter* or *Esc* key is pressed.

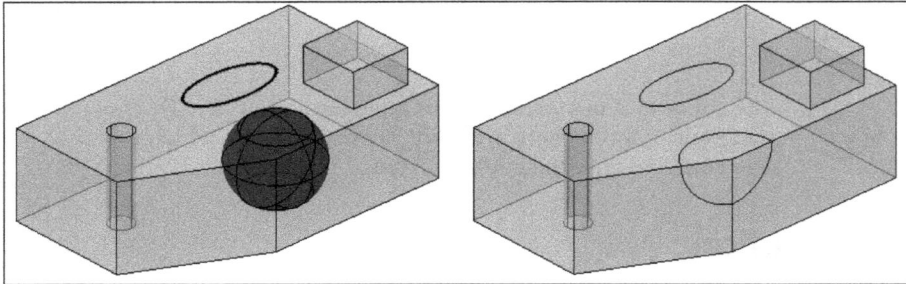

- **seParate solids**: If one solid has separated volumes, this option allows you to separate these volumes into independent solids. It prompts only for a solid selection.

- **Shell**: This option allows you to create a hollow solid with a constant wall thickness, eventually open through one or more faces. We can only apply this option once to a solid. After the solid selection, it prompts for the selection of faces to remove, thus creating an open thin-wall solid, and the shell thickness. A positive thickness creates the shell on the inside, while a negative thickness creates the shell on the outside.

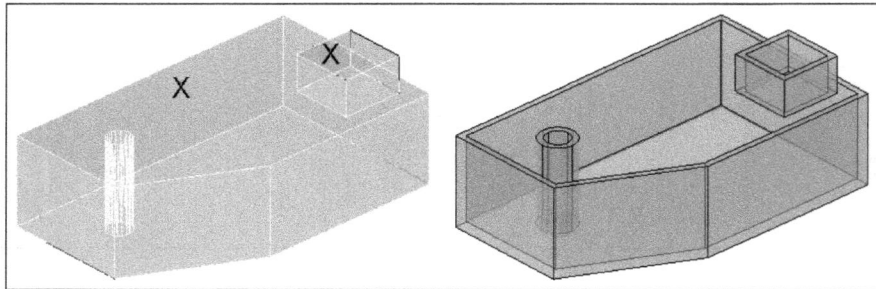

We must be careful with this option. Immediately after the solid selection, it prompts for face removal. If we press *Enter*, we remove no faces.

If we need some walls with a different thickness, after **Shell**, we may apply the **Move**, **Extrude**, or **Offset** face options.

- **Clean**: This option allows us to remove redundant edges or vertices, but does not remove edges coming from the **Imprint** option.

- **Check**: This option verifies whether the solid is valid. It prompts only for a solid selection.

Exercise 7.5

We are creating a building from exterior lines, mainly by applying the SOLIDEDIT command. It is advisable to use the **Solid Editing** toolbar or panel (ribbon) to access all SOLIDEDIT commands.

1. Open the drawing A3D_07_05.DWG. This drawing already has the 2D building contour.

2. Create and activate the layer 3D-BUILDING, color at choice.

3. Using PRESSPULL, click inside the area and specify 24 as the height of extrusion and end the command. We got the volume of our building.

4. To create the first slab and roof, we copy the horizontal faces. We apply the **Face/Copy** option of SOLIDEDIT, select the top and bottom faces, and specify a displacement of 40 along the X axis.

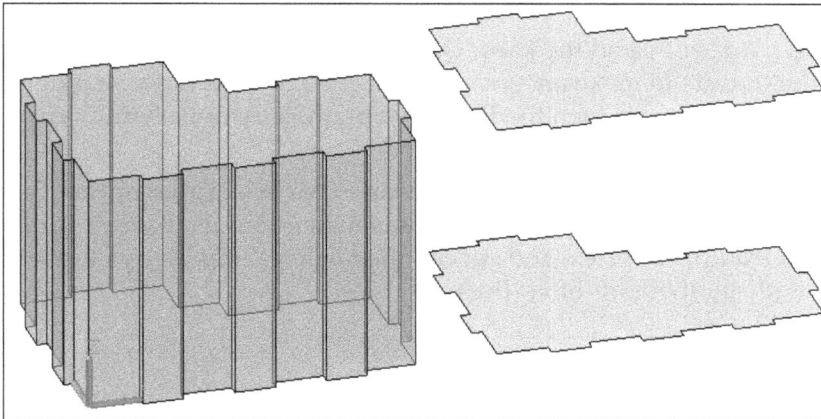

> By using a non wireframe visual style, it is easy to select the top face, orbit, and select bottom face. If we accidentally select vertical faces, we may use the **Remove** option.

5. Create two layers: 3D-SLAB and 3D-ROOF, colors at choice. Activate the first one.

6. Apply the EXTRUDE command to the lower copied face, with height -0.3.

7. Activate the 3D-ROOF layer and apply the EXTRUDE command to the upper copied face, with height 15.

8. With SOLIDEDIT and the **Body/Shell** option, make the main solid hollow. Select this solid, remove both horizontal faces, and specify thickness 0.3:

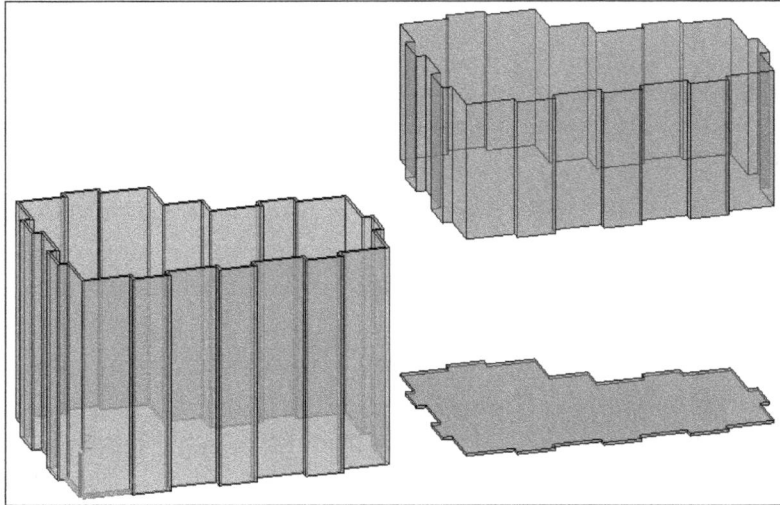

9. With SOLIDEDIT and the **Face/Offset** option, offset all the roof vertical faces to the outside to make an eave. To easily select these faces, mark one, apply the **All** option, and then the **Remove** option, to remove both horizontal faces. Specify the offset value 0.4.

10. To finish the roof, apply SOLIDEDIT and the **Face/Taper** option. Select again all vertical faces; the taper axis is defined from a bottom vertex to the projected upper vertex, and specify the angle 45. If it gives a modeling error, try with another pair of vertices.

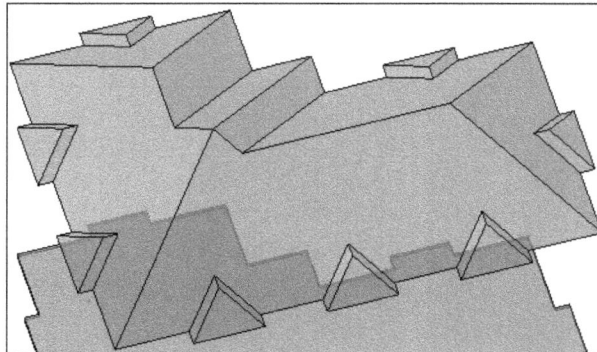

11. Make the roof hollow with SOLIDEDIT and the **Body/Shell** option. After selecting the roof, remove the bottom face and press *Enter*.

12. Move the roof and slab back to its position, that is, a displacement of 40 along the negative X axis.

13. The other slabs are shorter, so create one copy of the existing slab, shorten it by offsetting the vertical faces, and create the remaining copies. Now, copy the slab 3 units along the Z direction.

14. Isolate this slab (select it, right-click and choose **Isolate | Isolate Objects**). Then apply SOLIDEDIT and the **Face/Offset** option, offset all the slab vertical faces with the value -0.3. End isolation.

15. Finally, copy this slab seven more times with the same distance, 3, along the Z direction. This is the easiest way to apply the COPY command, **Array** option, eight elements, and distance 3.

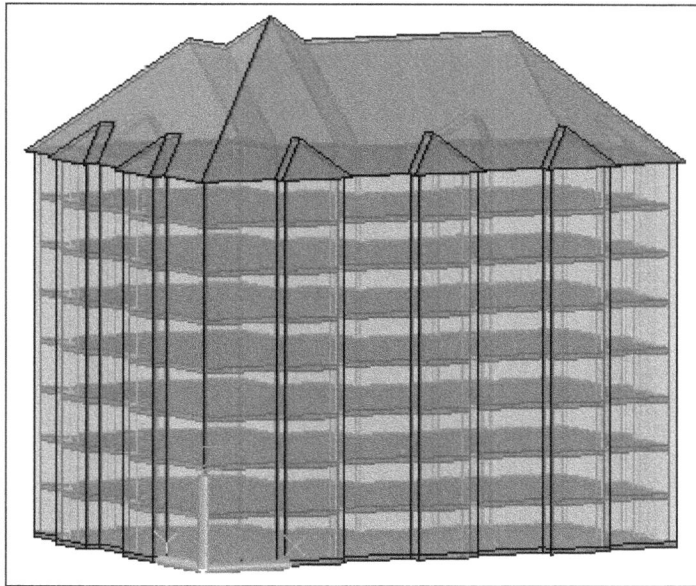

16. Save the drawing with the name A3D_07_05FINAL.DWG.

Creating linear objects from 3D

The following are two commands that create linear objects from solids or surfaces.

The OFFSETEDGE command

The OFFSETEDGE command (no alias), available since Version 2012, allows us to create a polyline or a spline parallel to the edge of a planar face of solids or surfaces. By default, the command prompts for the face and a point where it will pass the new linear object.

The command displays the default corner value and prompts for a planar face selection:

```
Command: OFFSETEDGE
Corner = Sharp
Select face: Selection
```

Then, by default, it prompts for a point where the new linear object will pass. It can be outwards or inwards, related to the face's edges:

```
Specify through point or [Distance/Corner]: Point
```

The face selection and point prompts are repeated until the *Enter* or *Esc* key is pressed:

```
Select face: Enter
```

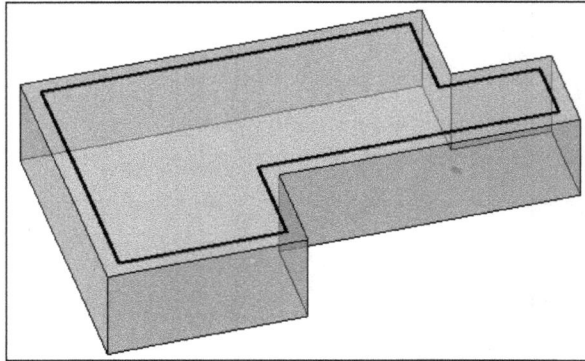

This command has the following options:

- **Distance**: Instead of specifying a point, this option prompts for a distance and to which side the offset is made.
- **Corner**: When offset is outside the edge, this option controls if the resulting corners are sharp or rounded. In this case, the radius is equal to the distance.

The SURFEXTRACTCURVE command

The SURFEXTRACTCURVE command (no alias), which is new in Version 2013, allows us to obtain curves (isolines) from surfaces or solids. The new objects can be lines, arcs, polylines, or splines. The command starts by displaying the default value of the **Chain** option and prompts for the selection of one surface, solid, or face:

```
Command: SURFEXTRACTCURVE
Chain = No
Select a surface, solid, or face: Selection
```

Specifying a point over the surface creates a curve. The prompt is repeated until the command is ended:

```
Extracting isoline curve in U direction
Select point on surface or [Chain/Direction/Spline points]: Point
Extracting isoline curve in U direction
Select point on surface or [Chain/Direction/Spline points]: Enter
```

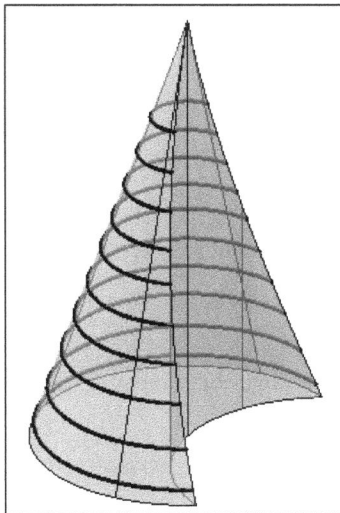

This command has the following options:

- **Chain**: With this option on, the isolines are also obtained from adjacent faces in the same direction.

- **Direction**: This option allows us to change the direction of isolines, between U and V.

- **Spline points**: This option allows marking points on the surface and creating a curve that passes through all points. This curve does not respect the U or V direction.

> In this version, the command has some weaknesses as it doesn't allow the specifying of points by coordinates.

Projecting linear objects

These two commands allow creating edges on solids or surfaces.

The PROJECTGEOMETRY command

The PROJECTGEOMETRY command (no alias), available since Version 2011, allows you to project points and linear objects on solids and surfaces. This projection may cut the solid or surface. The command starts by displaying the value of the SURFACEAUTOTRIM variable and prompts for the selection of curves or points to a project:

```
Command: PROJECTGEOMETRY
SurfaceAutoTrim = 0
Select curves, points to be projected or [PROjection direction]:
Selection
```

After the selection of objects to be projected, we select one solid, surface, or region:

```
Select a solid, surface, or region for the target of the projection:
Selection
```

Finally, we specify the projection direction between the viewing direction, the current UCS, or by marking two points:

```
Specify the projection direction [View/Ucs/Points] <View>: Option
```

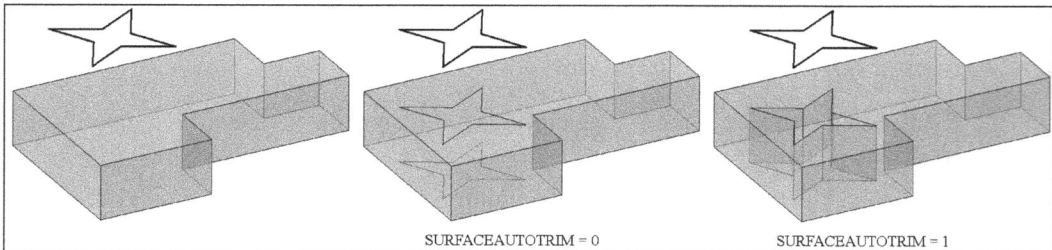

SURFACEAUTOTRIM = 0   SURFACEAUTOTRIM = 1

The command has the following option:

- **PROjection direction**: This option only allows us to specify the projection direction before selecting the objects to project.

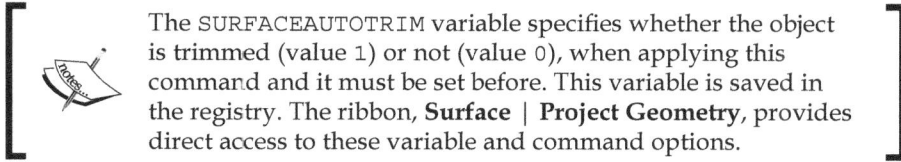

> The SURFACEAUTOTRIM variable specifies whether the object is trimmed (value 1) or not (value 0), when applying this command and it must be set before. This variable is saved in the registry. The ribbon, **Surface | Project Geometry**, provides direct access to these variable and command options.

The IMPRINT command

The IMPRINT command (no alias) allows you to create edges from intersections of solids and surfaces with linear entities, regions, or solids. This command is equivalent to the SOLIDEDIT **Imprint** option, the only difference being that IMPRINT can also be applied to surfaces.

Exercise 7.6

We are going to create a plastic part, with the application of the last seen commands.

1. Open the drawing A3D_07_06.DWG. This drawing is only composed by 2D polylines.
2. Start by orbiting the model and moving the inner polylines 30 units along the Z direction.
3. Apply the EXTRUDE command to the outer polyline, height of extrusion, 5.
4. With the FILLET command, round the bottom edges with a radius of 2. The **Chain** option can be used to select all edges.

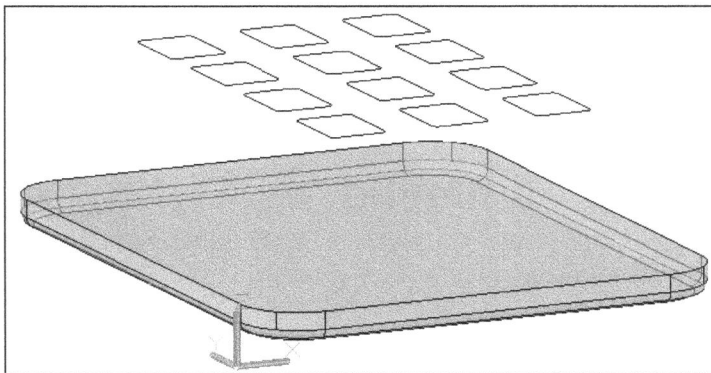

5. To create a small wall on the top part, start by applying the OFFSETEDGE command, selecting the top face, the **Distance** option, value 3, and a point on the inside. A polyline is created at a distance of 3 units from the edge.

6. Apply the PRESSPULL command to create the wall. We Specify a point between the edge and the polyline and height of extrusion, 10.

7. Before projecting the small polylines, you must verify that the SURFACEAUTOTRIM variable is set to 1.

8. Finally, apply the PROJECTGEOMETRY command, selecting all small polylines, then the solid and the **UCS** option. All polylines are projected onto the solid, cutting it.

9. We may erase or hide the small polylines.

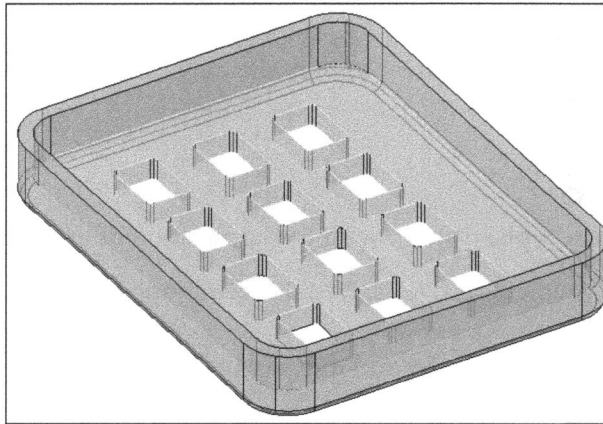

10. Save the drawing with the name A3D_07_06FINAL.DWG.

Editing surfaces

There is a command that is very useful when it comes to assigning thickness to surfaces.

The THICKEN command

The THICKEN command (no alias) allows us to apply a thickness to surfaces, thus creating a 3D solid. The command starts by prompting the surfaces to thicken:

```
Command: THICKEN
Select surfaces to thicken: Selection
```

We specify the thickness to be applied to all selected surfaces. It can be positive or negative:

```
Specify thickness <0.0000>: Value
```

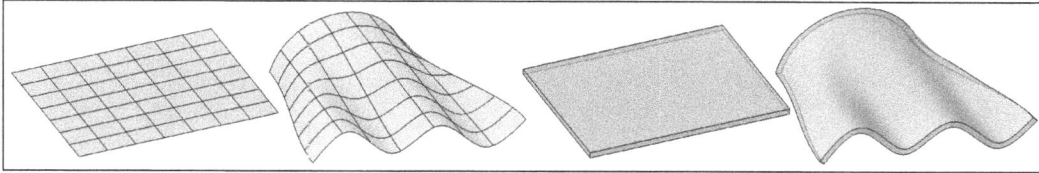

> If, when applying the command, there are self-intersections of volume, the operation may fail. If it fails, we may try a negative value.

Summary

We included in this chapter the main commands for editing solids and surfaces. We started by composing objects (UNION, SUBTRACT, INTERSECT) and editing them, then we presented the SLICE command to cut objects, the FILLET and CHAMFER commands (and the less used FILLETEDGE and CHAMFEREDGE commands) to fillet and chamfer objects. The PROPERTIES and QUICKPROPERTIES commands can be used to change creation parameters. Following was the SOLIDEDIT command that allows applying multiple operations on faces, edges, and volumes. To create linear objects from 3D, we have the OFFSETEDGE and SURFEXTRACTCURVE commands, and to project linear objects onto solids and surfaces, there are the PROJECTGEOMETRY and IMPRINT commands. Finally, we presented the THICKEN command to add thickness to surfaces.

In the next chapter, we will present all commands to inquire 3D objects.

8
Inquiring the 3D model

Frequently, we need to measure distances and volumes and obtain point coordinates. In 3D, it is also important to detect interferences and to obtain geometric properties of solids.

Topics covered in this chapter:

- How to obtain point coordinates
- How to measure distances and angles
- How to detect interferences
- How to create solids from interferences
- How to measure volumes
- How to obtain geometric properties

Measuring points, distances, and angles

Measuring and obtaining point coordinates is a constant part of our 3D work. The commands present here can be found on the **Inquiry** toolbar, and on the menu bar by going to **Tools | Inquiry**. Both **3D Basics** and **3D Modeling** workspaces don't have these commands, unlike the **Drafting & Annotation** workspace.

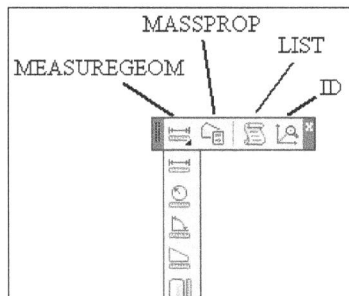

Obtaining point coordinates

One of the most important things in AutoCAD is to know precise point coordinates. For that, we apply the next command.

The ID command

The ID command allows us to obtain the X, Y, and Z coordinates of points related to the current **User Coordinate System (UCS)**. This command can be used transparently, that is, in the middle of another command, if we prefix it with a ' (single quotation) mark. The command only prompts for a point:

```
Command: ID
Specify point: Point
```

The command writes the X, Y, and Z coordinates, related to the current UCS, on the command line:

```
X = 28.9427     Y = 15.0000     Z = 10.0000
```

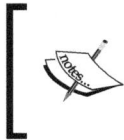

> If we want to get the coordinates of several points, we may use the MULTIPLE command instead of manually repeating the command. This command prompts for the command to repeat and ends when the *Esc* key is pressed.

Measuring distances and angles

To measure distances, perimeters, and angles, we can apply the following commands.

The DIST command

The DIST command (alias DI) is, for sure, one of the most commonly used commands, allowing us to measure distances, perimeters, and angles. By default, the command prompts for two points:

```
Command: DIST
Specify first point: Point
Specify second point or [Multiple points]: Point
```

The command returns the real distance (in 3D), the projected angle onto the XY plane, the angle from the XY plane, and all increments along the three directions:

```
Distance = 21.4783,  Angle in XY Plane = 29,  Angle from XY Plane = 16
Delta X = 17.9994,  Delta Y = 10.0666,   Delta Z = 6.0000
```

The result can be seen in the following screenshot:

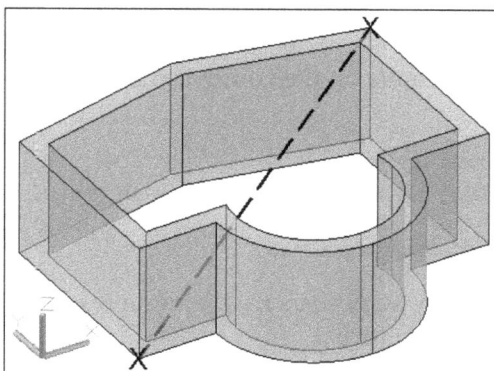

Since Version 2010, this command has one new option, which is as follows:

- **Multiple points**: This option allows you to specify multiple points, and returns the perimeter. We can specify a sequence of points and the command sums up the partial distances. Additional options allow the inclusion of arcs and the respective options, closing the perimeter, specifying a length, undoing the last point, and obtaining the total.

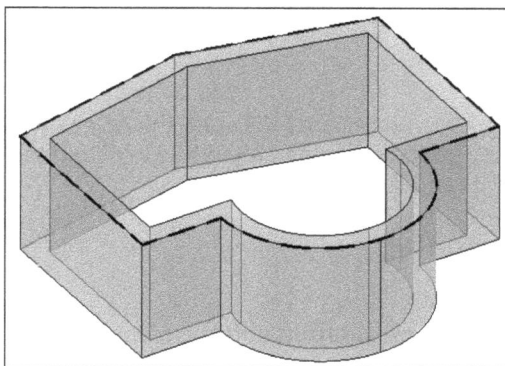

> The **Multiple points** options works the same way as the PLINE command. When including arcs, if they are not continuous, the best option is choosing the second point and specifying the middle object snap of that element.

The MEASUREGEOM command

The MEASUREGEOM command (alias MEA) allows us to obtain several measures, including distances, radius, angles, areas, and volumes. The command displays a set of options, shows the results on the command line, and repeats until we press the *Esc* key.

```
Command: MEASUREGEOM
Enter an option [Distance/Radius/Angle/ARea/Volume] <Distance>: Option
```

This command has the following options:

- **Distance**: This option works exactly the same way as the DIST command, which was presented before.

- **Radius**: This option prompts for the selection of an arc, a circle, or a circular edge and displays its radius and diameter.

- **Angle**: This option simulates the DIMANGULAR command, that is, it prompts for the selection of an arc, circle, line, or vertex and returns the angle. We can also select straight or circular edges.

- **ARea**: This option works exactly the same way as the AREA command.

- **Volume**: This option allows us to obtain volumes, similar to the **ARea** option. Volumes can be specified by a sequence of points (first the base perimeter and then the height) and an object selection. Additional options allow adding or subtracting volumes.

> As the AREA command (alias AA), the **ARea** option of the MEASUREGEOM command allows us to obtain the surface area of solids and surfaces.

Exercise 8.1

We are going to obtain some information about a 3D model.

1. We open the drawing A3D_08_01.DWG.

2. Applying the ID command, we obtain the coordinates of points A, B, C, and D:

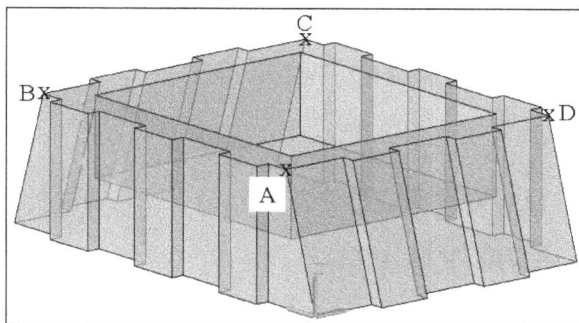

The coordinates of points A, B, C, and D are as follows:

```
X = 0.0000      Y = 3.0573      Z = 10.0000
X = 0.0000      Y = 28.9427     Z = 10.0000
X = 23.0000     Y = 28.9427     Z = 10.0000
X = 23.0000     Y = 3.0573      Z = 10.0000
```

3. Applying the DIST command and the **Multiple points** option, we obtain the perimeter of the top exterior border:

   ```
   Distance = 118.1363
   ```

   > The easiest way to obtain this value is to copy the top face (using the SOLIDEDIT command), explode the resulting region, and join all exterior segments into a polyline (using the JOIN command).

4. Applying the MEASUREGEOM command and the **Angle** option, we obtain the angle of the inclined faces:

   ```
   Angle = 73°
   ```

5. Again, with the last command and the **ARea/Object** option, we obtain the surface area of the solid:

   ```
   Area = 3142.9782
   ```

6. And without leaving the command, now with the **Volume/Object** option, we obtain the solid volume:

   ```
   Volume = 4808.5069
   ```

7. As there were no modifications, we don't need to save the drawing.

Interferences

When dealing with 3D objects, we often need to verify whether there are interferences.

Detecting interferences and creating solids with a common volume

The next command allows us to detect interferences between objects and to create new objects from common elements.

The INTERFERE command

The INTERFERE command (alias INF) allows us to detect interferences between two sets of solids or surfaces. If interferences are detected, the command allows the creation of solids, surfaces, or linear objects corresponding to the intersection, but the original objects are not affected. This command can be found in the **Solid Editing** panel on the **Home** tab in the ribbon, or on the menu bar by going to **Modify | 3D Operations**.

The command starts by prompting for the selection of the first set of objects, then a second set, and does a comparison between them. If we want to verify all objects, we just create one selection set. On using the command, it prompts for the first set. The *Enter* key is used to finish selecting:

```
Command: INTERFERE
Select first set of objects or [Nested selection/Settings]: Selection
```

It prompts for the second selection set or the *Enter* key that verifies all interferences between the objects in the first set:

```
Select second set of objects or [Nested selection/checK first set]
<checK>: Selection
```

Then, the interference verification starts and, on the first interference, the model is represented in wireframe, zoomed to the objects that interfere, the common element is represented in shaded red, and the **Interference Checking** dialog box is displayed. If there are no interferences, that information is displayed at the command line.

The **Interference Checking** dialog box has the following areas, buttons, and options:

- **Interfering objects**: It displays how many objects are present in each set and how many interfering pairs were detected.

- **Highlight**: The **Next** and **Previous** buttons show the next and previous interferences. With **Zoom to pair** checked, the corresponding interference is zoomed in.

- **The Zoom, Pan**, and **Orbit icons**: With the box visible, we can't navigate the scene. These buttons allow us to go to the model temporarily and zoom, pan, or orbit.

- **Delete interference objects created on Close**: With this option checked, the interference objects displayed are erased when leaving the command. If we want to maintain them, we uncheck it.

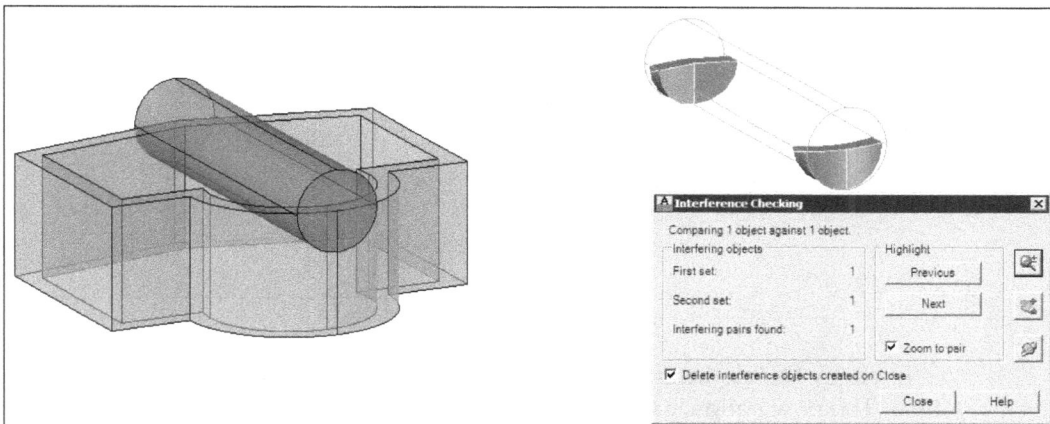

When selecting objects, there are two options available:

- **Nested selection:** This option allows us to include objects belonging to external references or blocks

- **Settings**: This option displays the **Interference Settings** dialog box, which allows us to configure the visual style and color applied to the interference objects, if the interference pair is highlighted and the visual style is applied to the remaining objects

Exercise 8.2

Let's check for some interferences. If there are any, we will create and subtract the interference volume.

1. Open the drawing A3D_08_2.DWG. This drawing contains several steel beams.

2. Use the INTERFERE command. The first selection set consists of two beams along the X direction and the second selection set consists of the remaining beams. The model changes to wireframe and the **Interference Checking** box is displayed, alerting us about ten interfering pairs.

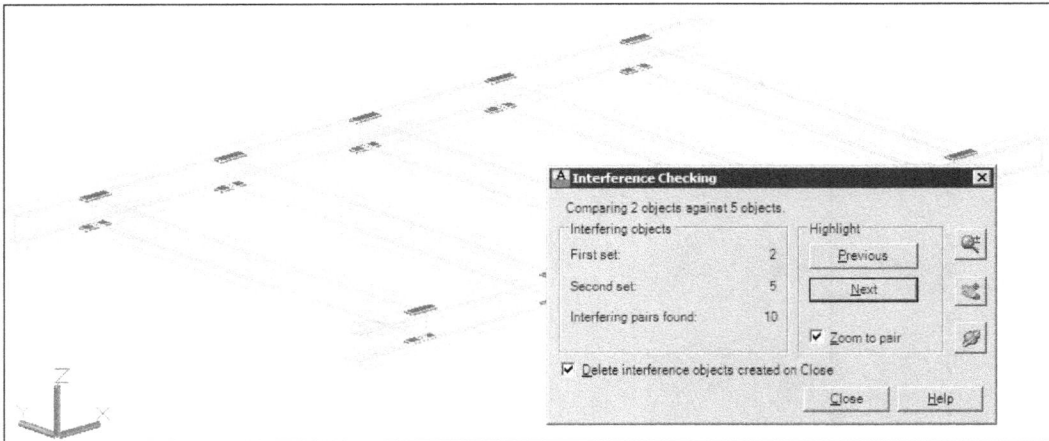

3. We uncheck **Delete interference objects created on Close** and click on **Close**. Ten new solids, each corresponding to an intersection, are created.

4. With the SUBTRACT command applied to each beam, we subtract the two corresponding intersection objects.

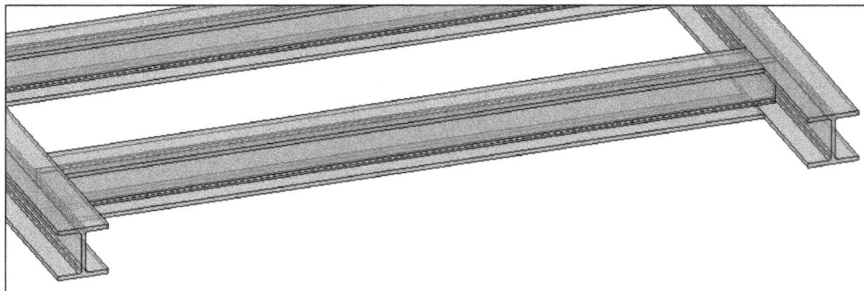

5. We save the drawing with the name A3D_08_02FINAL.DWG.

Volumes and other properties

During or after modeling, we may need to obtain geometric properties of solids, such as volumes for material acquisitions or moments of inertia for engineering calculations.

Obtaining volumes and other geometric properties

AutoCAD provides a very useful command for obtaining this type of information.

The MASSPROP command

The MASSPROP command allows you to obtain the most important geometric (or mass) properties for the selected solids. This command can be found on the **Inquiry** toolbar or the menu bar by going to **Tools | Inquiry**. It just prompts for the selection of solids and displays the respective properties:

```
Command: MASSPROP
Select objects: Selection
```

The properties displayed are as follows:

- **Volume and mass**: Volume is in cubic units. It also displays an equal value for mass, but is useless (it considers a density as 1).

```
---------------     SOLIDS     ---------------
Mass:    353.4154
Volume:  353.4154
```

- **Coordinates of the corners**: Coordinates of the opposite corners of the box that includes the selected solids related to the current UCS completely.

```
Bounding box: X: 800.6427  --  818.6427
              Y: 387.8904  --  404.8904
              Z: 0.0000    --  6.0000
```

- **Centroid or mass center**: It is related to the current UCS.

```
Centroid: X: 809.6336
          Y: 396.9529
          Z: 3.0000
```

- **Mass moments of inertia related to the current UCS**: These properties are used to determine the resistance to acceleration arising from rotation about an axis, and its unit is mass times the distance squared.

```
Moments of inertia: X: 55702738.4748
                    Y: 231683910.8130
                    Z: 287378167.3181
```

- **Mass products of inertia related to the current UCS**: These properties are used to calculate the balancing of the rotating bodies. Its unit is the same as the moment of inertia.

```
Products of inertia: XY: 113582913.8584
                     YZ: 420867.7892
                     ZX: 858410.9301
```

- **Radii of gyration**: It is another way to measure the moment of inertia and is also related to the current UCS. It is expressed in distance units.

```
Radii of gyration: X: 397.0046
                   Y: 809.6647
                   Z: 901.7458
```

- **Principal moments of inertia and respective directions**: These properties do not depend on the current UCS.

```
Principal moments and X-Y-Z directions about centroid:
I: 11330.2880 along [1.0000 0.0049 0.0000]
J: 14626.4757 along [-0.0049 1.0000 0.0000]
K: 23836.2712 along [0.0000 0.0000 1.0000]
```

Lastly, the command asks if this information is to be written to a text file with a .mpr extension:

```
Write analysis to a file? [Yes/No] <N>: Option
```

> Instead of creating an MPR file, it is faster to copy and paste this information from the text window (the *F2* key displays and removes it) to any program that includes a MTEXT object in AutoCAD.
>
> This command can also be used to calculate geometric properties of regions.

Exercise 8.3

We create a simple model composed of only a torus, to verify their mass properties.

1. We open AutoCAD or initiate a drawing.

2. We create a torus having its center at 10,10, radius as 1, and tube radius as 0.2, and zoom to it.

3. We apply the MASSPROP command and get the following results:

```
---------------    SOLIDS    ---------------
Mass:                   0.7896
Volume:                 0.7896
Bounding box:      X: 8.8000   --  11.2000
                   Y: 8.8000   --  11.2000
                   Z: -0.2000  --   0.2000
Centroid:          X: 10.0000
                   Y: 10.0000
                   Z: 0.0000
Moments of inertia: X: 79.3714
                    Y: 79.3714
                    Z: 158.7269
Products of inertia: XY: 78.9568
                     YZ: 0.0000
                     ZX: 0.0000
Radii of gyration:  X: 10.0262
                    Y: 10.0262
                    Z: 14.1785
Principal moments and X-Y-Z directions about centroid:
Press ENTER to continue:
                    I: 0.4145 along [1.0000 0.0000 0.0000]
                    J: 0.4145 along [0.0000 1.0000 0.0000]
                    K: 0.8133 along [0.0000 0.0000 1.0000]
```

4. Analyzing the results, we can confirm that the centroid position is the torus center. The principal moment of inertia about the Z axis is greater than the others, meaning that we need more force to rotate the solid about the Z axis than about the others.

5. We don't need to save the drawing.

Summary

This chapter includes all the commands used to obtain information from 3D objects, namely, ID to get point coordinates and DIST and MEASUREGEOM to get distances, perimeters, and other values. To check interferences, INTERFERE is available, which also allows creating objects corresponding to the common volume, surface, or lines. MASSPROP is the command used to obtain geometric or mass properties of solids.

In the next chapter, we will see how to obtain 2D drawings from 3D models.

9
Documenting a 3D Model

After creating a 3D model, we frequently have to create construction or fabrication drawings. First we are going to see how to define a layout and then the most important commands and procedures to obtain automatic 2D drawings from 3D models.

Topics covered in this chapter are, how to:

- Obtain 2D drawings from 3D models
- Configure and print layouts
- Create model sections
- Create flattened projected views
- Create associative views that update automatically
- Create associative sections and details

Creating 2D drawings from 3D models

AutoCAD provides several methods and commands for creating 2D drawings from 3D models. Along AutoCAD versions, this is a subject which is often improved. In this chapter, we will discuss the commands that simplify the creation of 2D drawings such as projections, sections, and details.

As all these drawings are normally intended to print, they should be placed on layouts. We start this chapter by reviewing how to create and configure layouts and viewports.

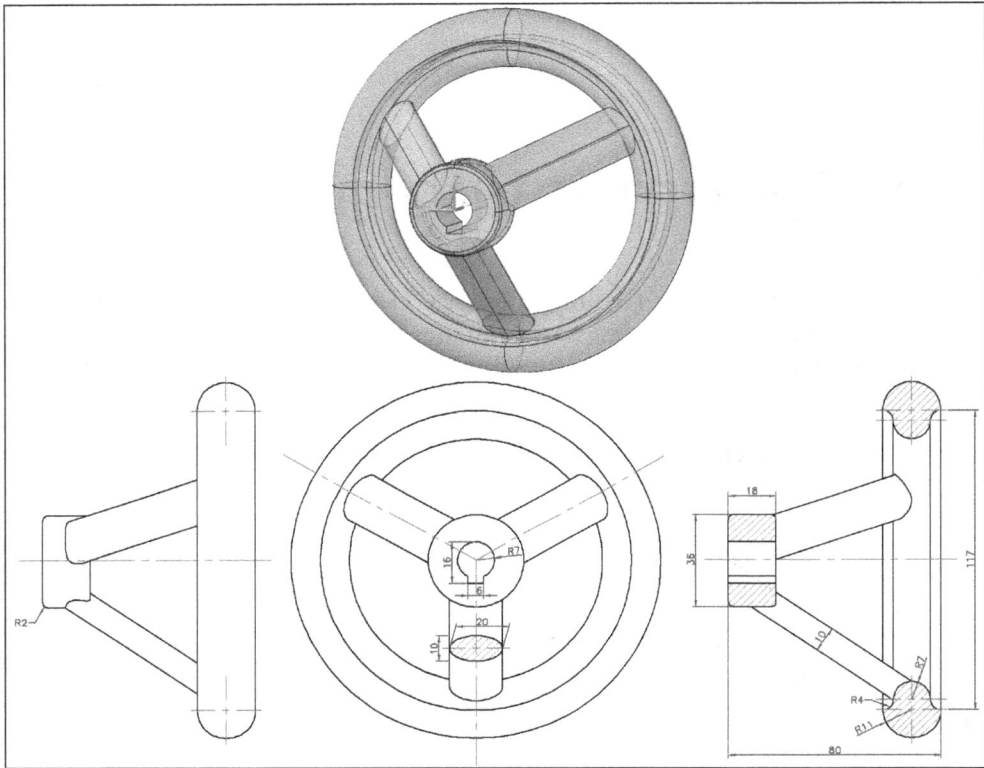

Layouts

A layout is the simulation of a paper sheet where we place viewports (windows to the model), blocks, text, frames, legends, and all useful elements to document the project.

We may have up to 255 layouts. By default, they are represented with a white background, the coordinate system icon is blue with a triangular shape. Also, the paper sheet and margins (dashed line) are displayed, and viewports may overlap or not fill the paper. Each viewport can have its own view, visual style, scale, and layer properties.

Following are the main steps to create a layout, configure it, place viewports and other elements, and control the layer properties.

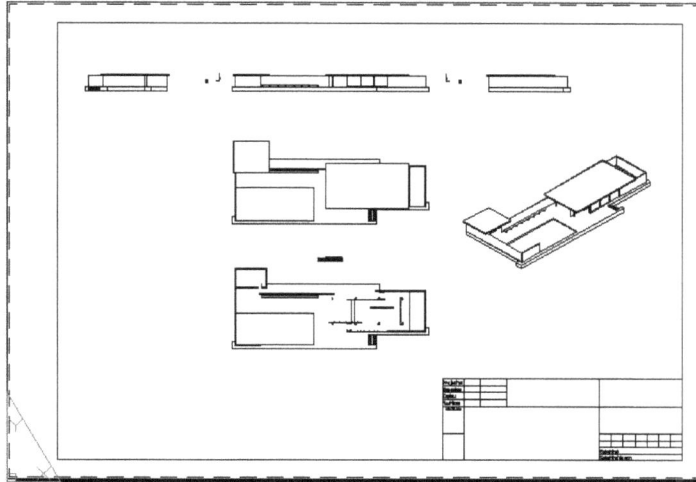

Creating and configuring layouts

We will demonstrate the process from scratch, that is, creating a new layout and then configuring it. All drawings must have at least one layout. By default, two layouts are available; which are **Layout1** and **Layout2**.

The first time we access a layout, a viewport is automatically created. Unless we maintain the default paper size, this viewport is useless and should be deleted. This behavior can be changed by unchecking the **Create viewport in new layout** option on the **Display** tab in the OPTIONS command.

Shortcut menu

The easiest way to create a new layout is by applying the shortcut menu (also known as contextual menus, right-click on the mouse button) with the cursor over the **Model** or any layout tab (just below the drawing area), and choosing **New Layout**. A new layout tab is created, named **Layout** with the lowest number available.

If our AutoCAD configuration has no tabs, there are two icons on the right-hand side of the status bar, which allow us to access **Model** or the last used layout. To restore tabs display, just right-click on any of these two icons and choose **Display Layout and Model Tabs** on the shortcut menu.

From this shortcut menu, we can also copy a layout from a template, delete, move or copy, rename, and select all layouts.

> To rename a layout, the best way is to double-click on the tab and write the new name. By dragging tabs, we can order layouts.

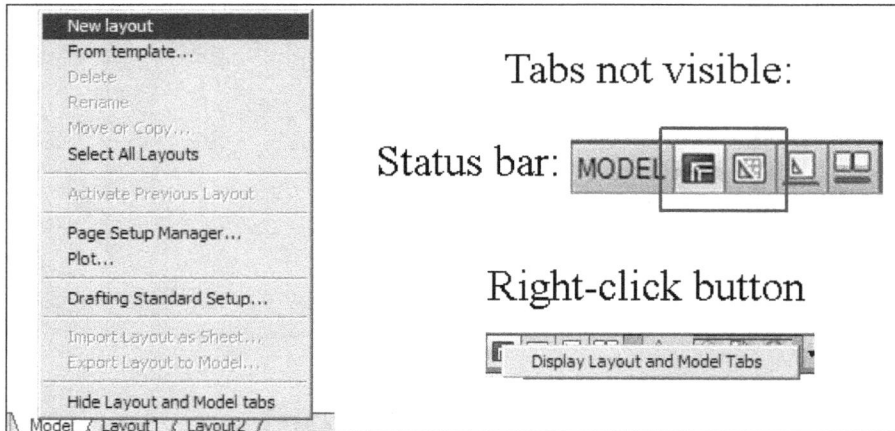

The LAYOUT command

Another way to create layouts is by applying the LAYOUT command. This command allows for several operations with layouts:

```
Command: LAYOUT
```

By default, the command prompts for an existing layout to activate:

```
Enter layout option [Copy/Delete/New/Template/Rename/SAveas/Set/?]
<set>:Option
Enter layout to make current <Layout1>:Name
```

The remaining options are:

- **Copy**: This option creates a new layout by copying an existing one
- **Delete**: This option deletes a layout whose name is specified
- **New**: This option creates a layout with a specified name
- **Template**: This option imports layouts from a DWG, DWT, or DXF file
- **Rename**: This option renames the specified layout
- **Saveas**: This option saves the specified layout to a new DWT file (template)
- **?**: This option lists all layouts

The PAGESETUP command

After creating a layout we must configure it with the PAGESETUP command. The configuration relates to the paper itself and has nothing to do with viewports or other graphical information. We can also access this command via the tab's shortcut menu and specify **Page Setup Manager**.

The command displays the **Page Setup Manager** dialog box where the saved page setups are listed. We may create a new page setup (the **New** button), modify the selected one (the **Modify** button), activate a page setup (the **Set Current** button), or import a page setup from another drawing (the **Import** button). With **Display when creating a new layout** checked, this box is displayed whenever we access or create a new layout:

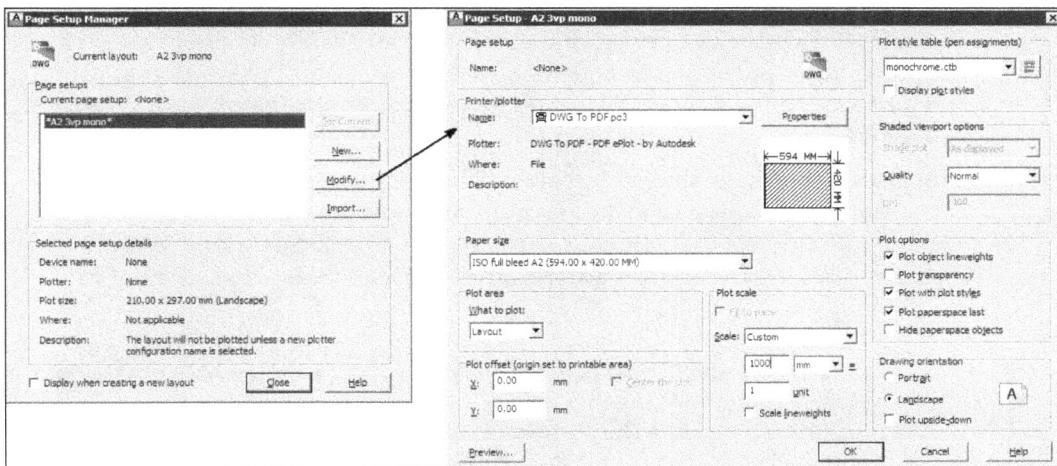

On selecting **New** or **Modify**, a new box is displayed.

> Normally, there is no need to define different page setups in the same layout. If we want a different configuration, it is better to create a new layout.

Everything that we need to configure the layout is in this box:

- **Printer/plotter**: In the **Name** list we choose the system printer (eventually PDF driver or other) or PC3 file, all are recognized by AutoCAD. System printers are represented by a small printer icon, while PC3 files are represented by a plotter icon. The **Properties** button displays **Plotter Configuration Editor**, where we can change the plotter/printer/driver configuration, for instance, resolution printing for images or setting up new paper sizes.

- **Paper size**: This list contains all paper sizes acceptable by the chosen printer or plotter.

- **Plot area**: We specify the area to plot. On layouts, we should choose **Layout**. When selecting **Window**, a button is displayed to define the area.

- **Plot offset (origin set to printable area)**: We can specify an offset for plot origin, along X and Y paper coordinates, with an additional option to center the plot area to the page. This option is not available if the plot area is set to **Layout**.

- **Plot scale**: We define the print scale by choosing a standard scale or by corresponding a value in inches or millimeters to a value in drawing units. **Fit to paper** adjusts the plot area to the page size and **Scale lineweights** changes drawing linewidths according to the scale (it is advisable not to do this).

> If we are simulating the paper sheet, the plot scale should be 1:1 (1 inch or mm to 1 drawing unit). But if our drawing unit is other than inches or mm? The best way is to specify at the box near the unit (inches or mm), our drawing unit is converted to that unit. For instance, if our model unit is in meters, we should specify 1000 mm. So our real scale is 1000:1, telling AutoCAD that plot scale is 1:1 but our drawing unit is in meters.

- **Plot style table (pen assignments)**: We specify a plot style to the layout, which, among other properties, defines the printing colors and lineweights. By pressing the button to the right, we can verify the chosen plot style, eventually making changes and creating a new one. Checking **Display plot styles**, plot styles are applied directly to the layout.

- **Shaded viewport options**: Within this area, it is possible to print viewports with shaded or rendered views. On layouts, **Shade plot** is not available, it is a viewport property.

- **Plot options**: This area includes some additional options, namely, if we want to print with object lineweights, transparency, and plot styles.

- **Drawing orientation**: We choose between portrait or landscape drawing orientation.

- **Preview**: This button allows us to preview the print.

Creating viewports and adjusting scales

After configuring page setup, we need to create viewports, and adjust their views and scales.

The MVIEW command

The MVIEW command (alias MV) allows creating viewports and includes some related options. This command can also be found on the **Viewports** toolbar, on the ribbon, in the **Layout** tab, and in the **View/Viewports** menu bar. When starting the command, it displays several options, the default one allows to create a new rectangular viewport by specifying two opposite corners:

```
Command: MVIEW
Specify corner of viewport or [ON/OFF/Fit/Shadeplot/Lock/Object/
Polygonal/Restore/LAyer/2/3/4] <Fit>: Point
Specify opposite corner: Point
```

Viewports are created in the current layer. After creating a viewport, the model is represented with the same viewing direction and visual style, and zoomed to fit the viewport. Obviously, the scale is not normalized but this will be set up later.

Options for this command are:

- **ON/OFF**: To the selected viewports, **ON** displays the model, while **OFF** turns it off.

- **Fit**: This option creates a rectangular viewport that fits the printable area.

- **Shadeplot**: This option allows to specify the visual style or render view to the selected viewports, the default being **As Displayed**.

- **Lock**: If we access the model inside the viewport and zoom in or out, the scale factor is modified. To avoid this, we may lock the selected viewports with this option.

- **Object**: This option prompts to select a closed object and uses it as the boundary of a new viewport. We may select a circle, ellipse, region, closed polyline, or closed spline.

- **Polygonal**: This option creates a polygonal viewport by specifying several vertices. We can apply some options similar to the PLINE command, such as **Arc** to define arc segments.

- **Restore**: This option creates viewports based on a viewport configuration saved with the VPORTS command.

- **Layer**: This option allows to remove all layer property overrides for the selected viewports.

- **2/3/4**: This option creates 2, 3, or 4 viewports, by defining the distribution and two opposite corners of a rectangle that limits these viewports.

The **Lock** and **Shadeplot** options may also be changed anytime with the **Properties** palette:

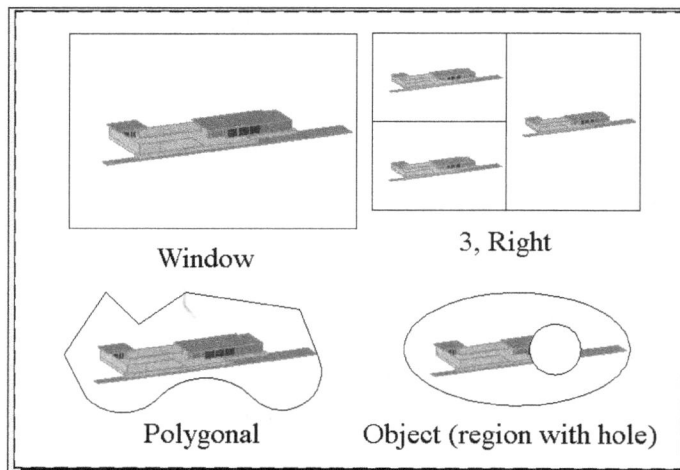

Window

3, Right

Polygonal

Object (region with hole)

Adjusting views

To adjust what we see in each viewport, we must go inside, that is, we access model space by maintaining the layout configuration. This intermediate state is called floating model space. Some visual changes occur: the cursor appears only inside the active viewport, elsewhere is represented by an arrow; the active viewport boundary is thickened, and the triangular coordinate system icon disappears. To access this state, there are three main options:

- Double-click inside the viewport, this is the easiest and recommended method
- Apply the MSPACE command, this command (alias MS) activates the last accessed viewport; if no accesses occur, it activates the last viewport created
- By pressing the button named **Paper/Model**, on the status bar, we access the MSPACE command

Once inside a viewport, we can zoom, pan, orbit, or apply an orthogonal or saved model view. We should specify and center the view in each viewport. Also, at this moment, it is possible to adjust layer properties specific to each viewport (presented in the next section).

A single click on another viewport activates it. To return to paper space, we double-click outside the viewports, apply the PSPACE command (alias PS), or press the **MODEL** button on the status bar.

Viewport scales

After defining the model view in all viewports, we arrive at one of the most important steps: adjust viewport scales. While creating technical drawings for manufacturing, construction, or other areas, most viewports must have a technical normalized scale. As usual in AutoCAD, there are several methods available:

- **Scale list on the status bar**: We select one or more viewports and pick one of the predefined scales from this list. Advantages of this process: it is always available when we select a viewport; it is easier when using the annotative property of objects (beyond the scope of this book) and provides access to the SCALELISTEDIT command (**Custom** option). This method doesn't allow for writing scales.
- **Scale list on the Viewports toolbar**: We select one or more viewports and pick from the list of this toolbar. If the desired scale is not listed, we may write it, in the format **1:200** or **1/200**.
- **The PROPERTIES command**: In the **Properties** palette, from the selected viewports, we may choose a predefined scale, in **Standard scale**; or write it using the same formats as before, in **Custom scale**.

- **Apply the** ZOOM **command with value**: We enter the viewport, apply the ZOOM command, and write the scale value with the suffix **XP**, for instance, **1/200XP** or **0.005XP**:

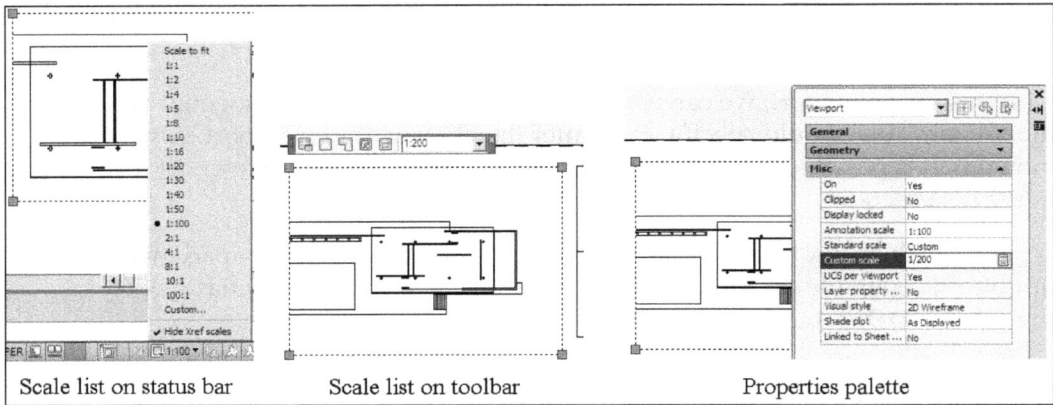

| Scale list on status bar | Scale list on toolbar | Properties palette |

> I advise applying the scale list on the status bar. Normally this list includes all the needed scales, according to the following command.

The SCALELISTEDIT command

The SCALELISTEDIT command (no alias) defines the list of scales available in the current drawing. We can access this command by the scales list on the status bar and select **Custom**. The command displays a dialog box with a scales list and some buttons:

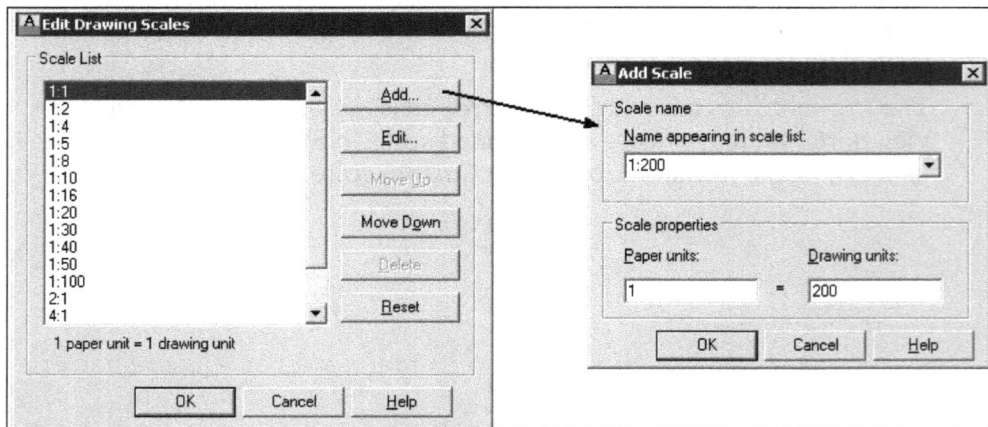

The functionalities of the buttons are as follows:

- **Add**: This button displays a new box allowing to add a new scale. We specify the name appearing on the list and the correspondence between paper units and drawing units.

- **Edit**: This button allows you to modify a scale, displaying a box similar to **Add**. If the scale is already applied to viewports or annotative objects, it cannot be modified.

- **Move Up / Move Down**: These options allow reordering of the scales list.

- **Delete**: This button deletes the scale. If the button is grayed out, it means that it is being used and cannot be deleted.

- **Reset**: This button inserts the default scales saved with the program in the drawing. Any unused scale not belonging to the default scale list is deleted. The default scale list is defined with the OPTIONS command, **User Preferences** tab, and **Default Scale List** button. We may specify metric or imperial scales lists.

> There is also a command-line version, -SCALELISTEDIT (name preceded by a minus sign), useful for creating scripts and automate processes.

Plot styles

Plot styles specify printing colors, lineweights, grayscale, screening (color intensity), and other properties. When creating a drawing, this is defined with one of the two possible types of plot styles:

- **Color plot style (CTB file)**: The main 255 colors (**ACI** that stands for **AutoCAD Color Index**) control printing properties, each with specific properties. All objects sharing the same color (explicit or by layer) will be printed with the same properties.

- **Named plot style (STB file)**: Instead of 255 colors, names are defined in each STB file, and properties (identical to CTB) are assigned to each name. Commonly, these names are assigned to layers. For instance, we may have a layer called **Walls**, with a plot style name **0.3BLACK** assigned to it.

The CONVERTPSTYLES command (no alias) allows to convert a drawing from color plot style to named plot style and vice versa.

Layer properties per viewport

One of the great advantages of applying layouts is the layer properties control for each viewport. We can freeze layers per viewport, and also modify colors, lineweights, linetypes, and other properties. To override properties per viewport we must first activate it.

Freezing layers per viewport

The layers list on the **Layers** toolbar or ribbon inside the **Layers** panel, in the **Home** tab allows freezing a layer only in the active viewport. We access this list and mark on the symbols of the third column, available only on layouts. Objects on these layers remain visible in other viewports and in the model space:

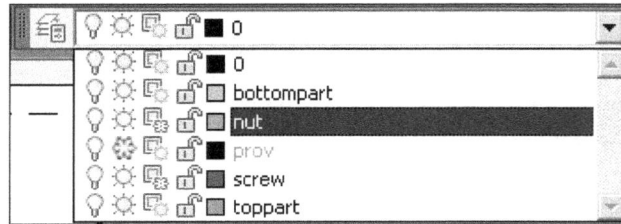

The LAYER command

The LAYER command (alias LA), has several new columns, when called on layouts. These columns allow you to specify different layer properties only in this viewport:

- **New VP Freeze**: When creating new viewports, visibility of existing layers is controlled by this column

- **VP Freeze**: This column allows freeze or thaw layers only in the current viewport, being the same as the third column of the layers' list

- **VP Color**: This column allows to modify the layers' colors only for the active viewport

- **VP Linetype**: This column allows to modify layers' line-types only for the active viewport

- **VP Lineweight**: This column allows to modify layers' lineweights only for the active viewport

- **VP Transparency**: This column allows to modify layers' transparencies only for the active viewport

- **VP Plot Style**: If the drawing plot style is STB based, it allows specifying names of the active plot style only for the active viewport:

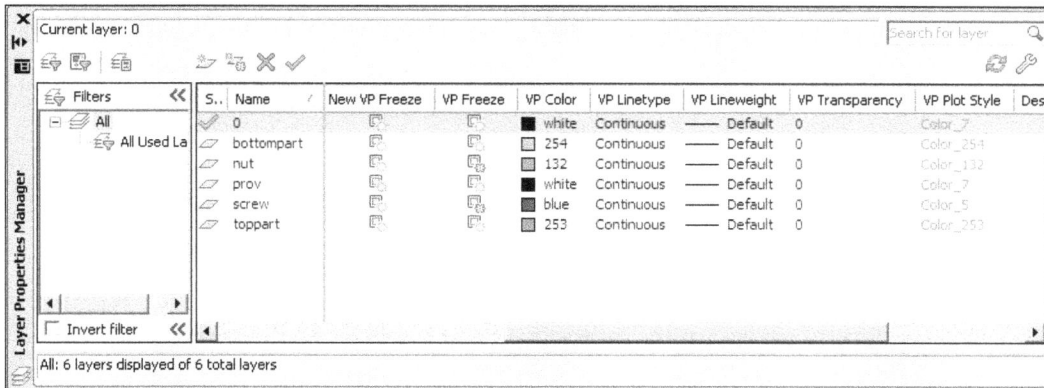

S..	Name	New VP Freeze	VP Freeze	VP Color	VP Linetype	VP Lineweight	VP Transparency	VP Plot Style	Des
	0			white	Continuous	Default	0	Color_7	
	bottompart			254	Continuous	Default	0	Color_254	
	nut			132	Continuous	Default	0	Color_132	
	prov			white	Continuous	Default	0	Color_7	
	screw			blue	Continuous	Default	0	Color_5	
	toppart			253	Continuous	Default	0	Color_253	

Current layer: 0

All: 6 layers displayed of 6 total layers

Previewing and printing

After specifying the layout, it is time for a last preview and to print.

The PREVIEW command

The PREVIEW command (no alias) displays a **WYSIWYG** (**What You See Is What You Get**) preview of printing.

The PLOT command

The PLOT command (shortcut *Ctrl + P*) displays the **Plot** dialog box. This box is similar to the **Page Setup** dialog box and contains all elements specified on the page configuration. The differences are: the **Number of copies** option, to print several copies of the layout; the **Plot stamp on** option, to place the defined plot stamp; and the **Save changes to layout** option, to apply any changes made on this box to page setup.

Exercise 9.1

From a 3D mechanical model in millimeter (mm), we are going to configure a layout and create four viewports. We want an A3 format sheet, printing everything back.

1. Open the drawing A3D_09_01.DWG.

2. A layout is going to be configured. Double-click on the **Layout1** tab to modify its name to **A3-Assembly-4vp**.

3. With the shortcut menu on this tab, choose the **Modify** option in the **Page Setup** manager. In printer/plotter select **DWG to PDF.pc3**; Paper size, **ISO full bleed A3 (420.00 x 297.00 MM)**; plot style tale, **monochrome.ctb**; and check the option **Display plot styles**. Press **OK** and **Close**.

4. If there is already a viewport, delete it. We create and activate a layer called **Viewports**, with the **No plot** property.

5. With the MVIEW command, create one rectangular viewport and copy it three times. Then, create a circle and, with the MVIEW command's **Object** option, define a viewport delimited by the circle:

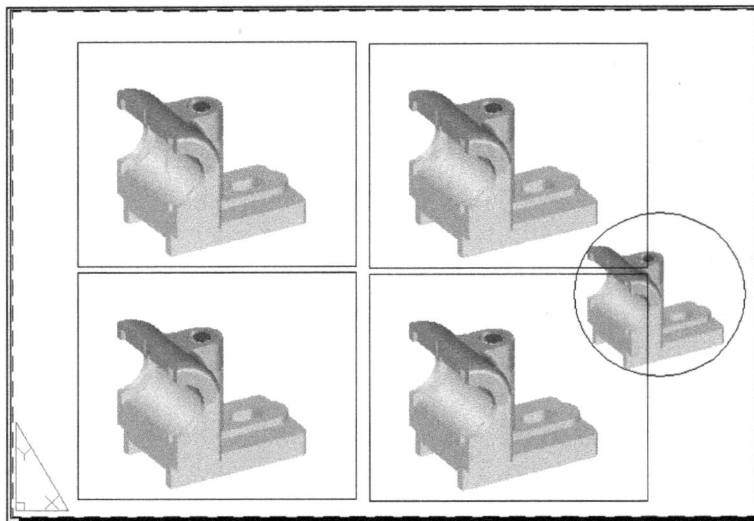

6. Now, we are going to define what to see inside each viewport. Double-click inside the top-left one, change the visual style to **2D Wireframe**, and specify a **Front** view (for instance, with the **ViewCube** tool).

7. Activate the top-right viewport with a single click, change the visual style to **2D Wireframe**, and specify a **Left** view. To the lower-left viewport, apply the same visual style and specify a **Top** view. To the lower-right viewport, again, apply the same visual style, but maintain a perspective view. The round viewport should also have the same visual style and a **Top** view. Double-click outside to return to paper space.

8. Let's specify scales for all viewports. Select all rectangular viewports and, on the scales list (status bar), select **1:1**. To the circular viewport, apply **2:1**:

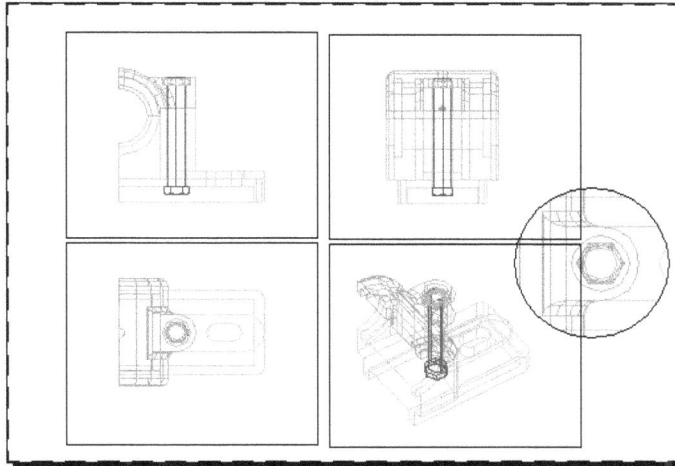

9. Orthographic views should be aligned and maybe we could reduce the viewport's size to better organize the layout.

10. Write, below each orthographic viewport, the respective scale. To complete the layout, insert additional elements (for instance, frame and drawing information).

The PREVIEW command should display a view similar to the next diagram:

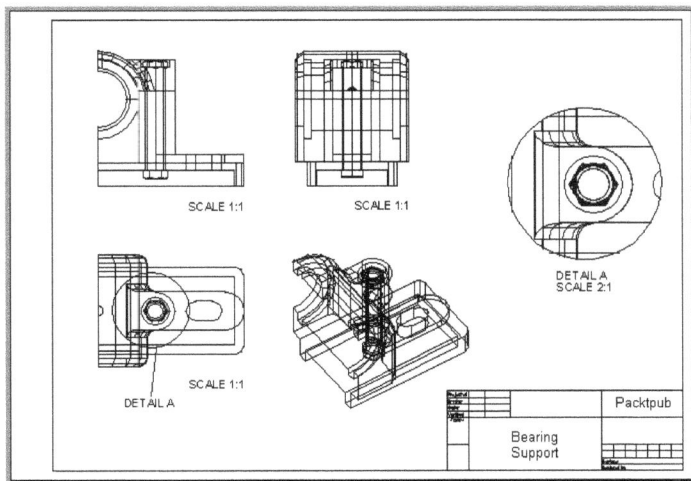

> From the previous exercise, we see that displaying orthographic views is far from perfect. If we apply a **3D Hidden** visual style to some views, the result is still not good. Next, we will see better alternatives.

Views and sections

Since for ever, documentation is a key part of 3D modeling. AutoCAD has several commands for creating projection, views, and sections from the model.

Sections

There are two commands for obtaining sections, the first one being more flexible and allowing the analyzation of 3D models.

The SECTIONPLANE command

The SECTIONPLANE command (alias SPLANE) allows you to specify visual section cuts, composed of one or more planes. The command can be accessed on the ribbon, **Home** tab, **Section** panel, or on the **Draw/Modeling** menu bar. It can section solids, surfaces, and meshes.

The command prompts for the selection of a face or by specifying a point that defines the beginning of the section plane:

```
Command: SECTIONPLANE
Select face or any point to locate section line or [Draw section/
Orthographic]: Face selection
```

After selecting the face, a plane is placed coincident with that face. This new object is called a section object, it will include all models and can then be activated and moved. If we specify a point not on any face, a second point is requested and a vertical plane is created that passes through both the points. We may have several planes defined, but only one can be activated.

Options for this command are:

- **Draw section**: We draw a section object composed of more than one plane. A sequence of points is requested and then, a point that defines the direction of the section view is drawn.

- **Orthographic**: We create an orthographic section plane, according to the active coordinate system. This option only prompts for the orthographic plane.

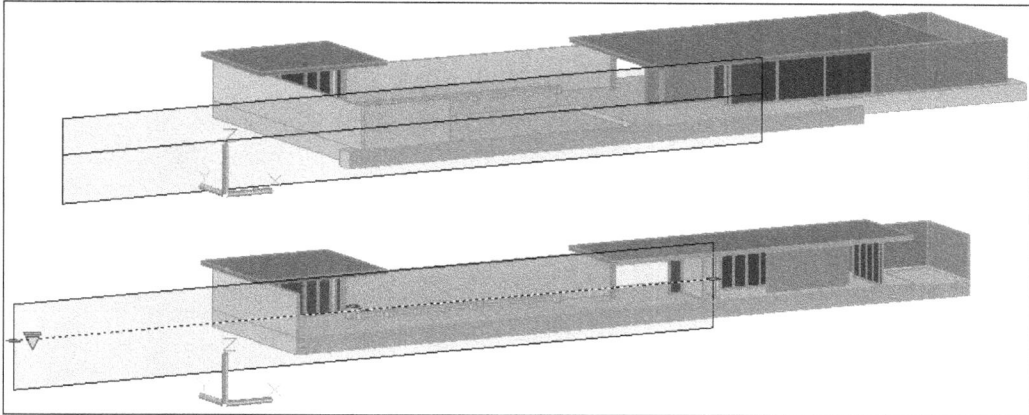

If we select a section plane and apply the shortcut menu, the access to several functions and related commands is provided:

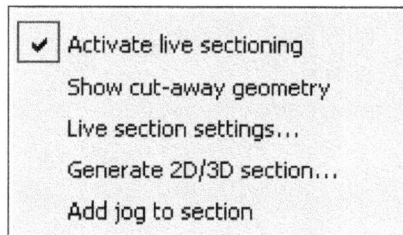

The functions and commands are listed as follows:

- **Activate live sectioning**: This applies the LIVESECTION command that activates or deactivates the section plane

- **Show cut-away geometry**: With this option checked, the cut geometry is represented with properties specified by settings; by default, red and transparent

- **Live section settings**: This applies the SECTIONPLANESETTINGS command. This command controls the section parameters

- **Generate 2D/3D section**: This applies the SECTIONPLANETOBLOCK command. This command displays a dialog box

- **Add jog to section**: This applies the SECTIONPLANEJOG command that allows adding vertices to the section line, thus creating a new plane with a 90 degree angle

Generating 2D/3D section

One of the major utilities of this command is the section generation. This section is generated as an anonymous block that can be placed, for instance, on a layout. By applying **Generate 2D/3D section** on the section plane shortcut menu (or the `SECTIONPLANETOBLOCK` command), a dialog box is displayed, with the following options:

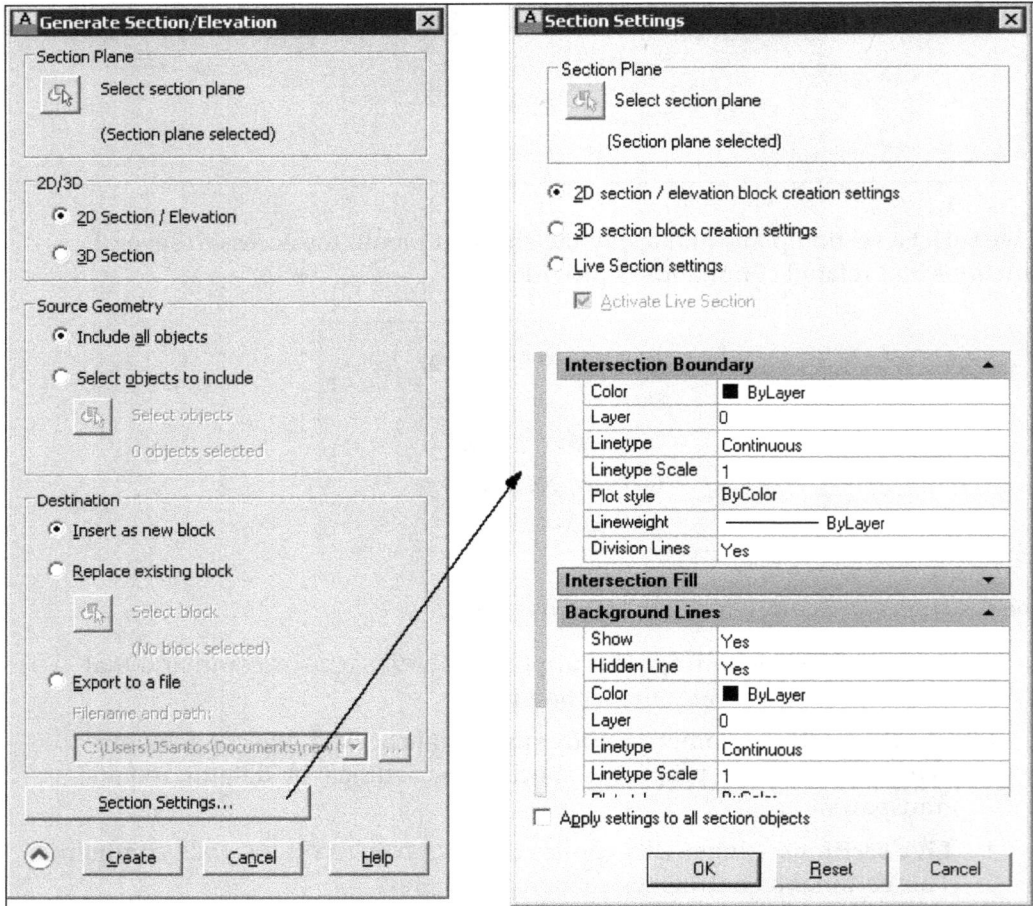

In **Section Plane**, we can select the plane. In **2D/3D**, we can choose between creating a planar 2D section or a 3D section. In **Source Geometry**, we can specify if the whole model is used for the section block generation, or if any specific objects are selected using the button below it. **Destination** specifies if the section is inserted in the drawing as a new block; it replaces an existing block, or is exported as a DWG or DXF file.

The button **Section Settings** calls the SECTIONPLANESETTINGS command, allowing you to specify all section parameters. These are divided by **2D section / elevation block creation**, **3D section block**, and **Live Section**. The 2D section contains all the following five panels:

- **Intersection Boundary**: This panel contains all parameters, such as color, layer, and lineweight, for lines corresponding to the intersection of geometry with the section plane, cut line
- **Intersection Fill**: This panel defines the hatch or fill corresponding to material cut by the section plane
- **Background Lines**: This panel defines all parameters for background lines; this panel is not available for **Live Section**
- **Cut-away Geometry**: This panel contains the parameters for cut-away geometry; by default, this is not shown
- **Curve Tangency Lines**: This panel controls the appearance of edges shared by curved and planar continuous faces, but this panel is only available for 2D sections

If we choose to insert a block, we press the **Create** button. A block is created and the command prompts for the insertion point, X scale factor, Y scale factor, and rotation. The block is inserted in the current UCS.

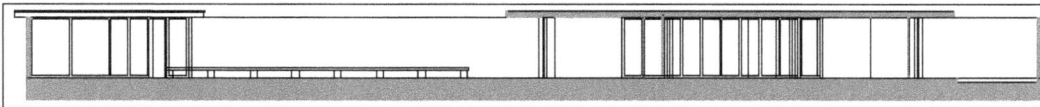

> These blocks are anonymous and may be exploded or edited with the BEDIT command.

The SECTION command

The SECTION command (alias SEC) is an old command for creating sections from 3D geometry, but very easy to apply. The result is a region or a curve for each cut object. The command prompts for the selection of 3D objects (can be solids, surfaces, or meshes):

```
Command: SECTION
Select objects: Selection
```

By default, the section plane is defined by three points:

```
Specify first point on Section plane by [Object/Zaxis/View/XY/YZ/
ZX/3points] <3points>: Point
Specify second point on plane: Point
Specify third point on plane: Point
```

The other options are:

- **Object**: The section plane is defined by a planar object to be selected, such as a circle, an ellipse, an arc, a 2D spline, or a 2D polyline

- **Zaxis**: The section plane is defined by a point on that plane and a second point that specifies the normal to the plane

- **View**: The section plane is parallel to the current view that passes through a point to be specified

- **XY/YZ/ZX**: The section plane is parallel to the XY, YZ, or ZX plane that passes through a point to be specified

Exercise 9.2

With the same model, we are going to create some sections:

1. Open the drawing A3D_09_2.DWG.

2. The sections should be on a proper layer, so create a **SECTIONS** layer and make it the current layer.

3. With the SECTION command create a section parallel to the ZX plane that passes through the middle of the assembly. The command creates four regions overlapping the geometry.

4. Move aside the four regions that are created.

5. Again with the SECTION command, create a section parallel to XY plane that passes through the center of the bolt, and move the four new regions to the side:

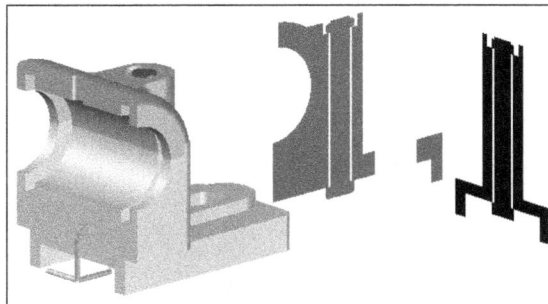

6. Create a new layer called **SECTIONPLANES** and make it current. Freeze the **SECTIONS** layer.

7. With the `SECTIONPLANE` command we want a section coincident with the ZX plane, for instance, applying the **Orthographic** option and, then the **Front** option. The plane cuts the assembly in half.

8. With the shortcut menu, activate **Show cut-away geometry**. The first half of the assembly is represented in red and transparent.

9. Let's now create a block that represents this section. Again, with the shortcut menu, choose **Generate 2D/3D section**. At the **Generate Section/Elevation** box, confirm **2D Section/Elevation**, **Include all objects**, and **Insert as new block**. We press the **Section Settings** button and modify the following: **Face Hatch** as **User-defined**; **Angle** as **45**; **Hatch Spacing** as **3**; **Color** as **Blue**; **Background Lines Hidden Line** as **No**; **Curve Tangency Lines Show** as **No**. Press **OK** and **Create**. The block is displayed, on the XY plane. Specify an insertion point, for instance, X scale 1, Y scale 1, and rotation 0.

10. Save the drawing with the name `A3D_09_2final.DWG`:

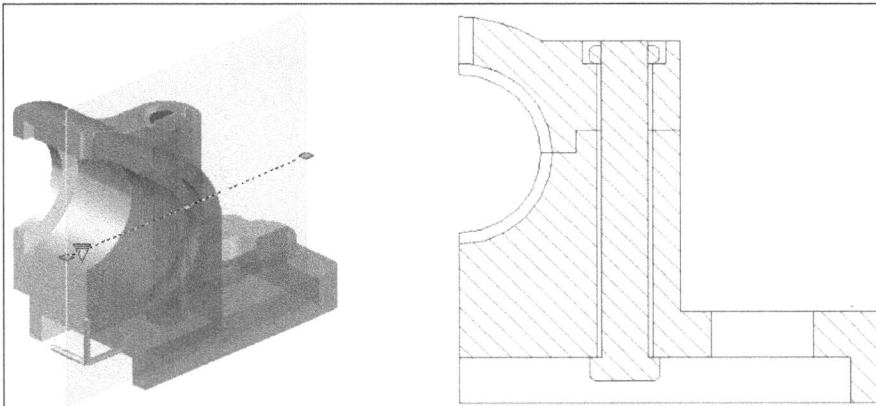

Projections and flattened views

While the `SECTIONPLANE` command creates sections, the next command creates projections, or flattened views.

The FLATSHOT command

The FLATSHOT command (no alias) creates a 2D representation of all 3D objects in the drawing by projecting all edges onto a plane perpendicular to the viewing direction. It can be applied to solids, surfaces, meshes, and regions.

First, we must set the viewing direction. Upon calling the command, a dialog box is displayed, including three areas and an option:

- **Destination**: This area is similar to the same area as the SECTIONPLANEBLOCK command and specifies whether the view is inserted in the drawing as a new block, it replaces an existing block, or is exported as a DWG or DXF file
- **Foreground lines**: This area allows you to specify the color and linetype for visible edges in the projected view
- **Obscured lines**: This area allows you to specify whether hidden edges are represented in the projected view and, in affirmative case, their colors and line-types
- **Include tangential edges**: When this option is checked, it draws profile or silhouette lines for curved surfaces or faces

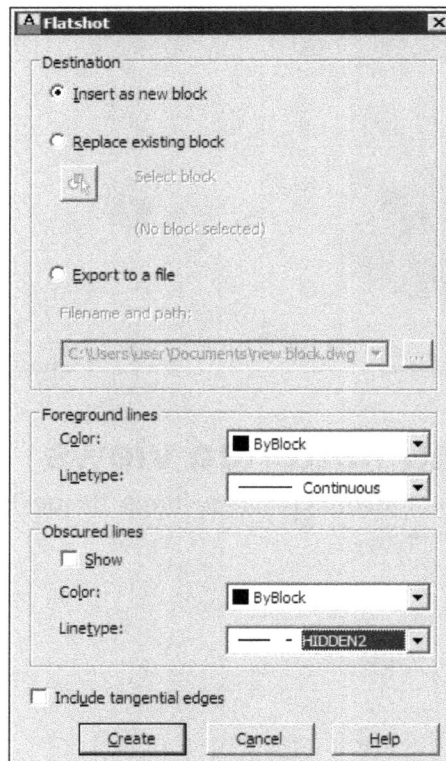

As before, if we choose to insert a block, when pressing the **Create** button, a block is created. The command prompts for insertion point, X scale factor, Y scale factor, and rotation. The block is inserted in the current UCS.

> These blocks are also anonymous and may be exploded or edited with the BEDIT command.
>
> The **ViewCube** tool is handy to specify orthographic views without creating coordinate systems.

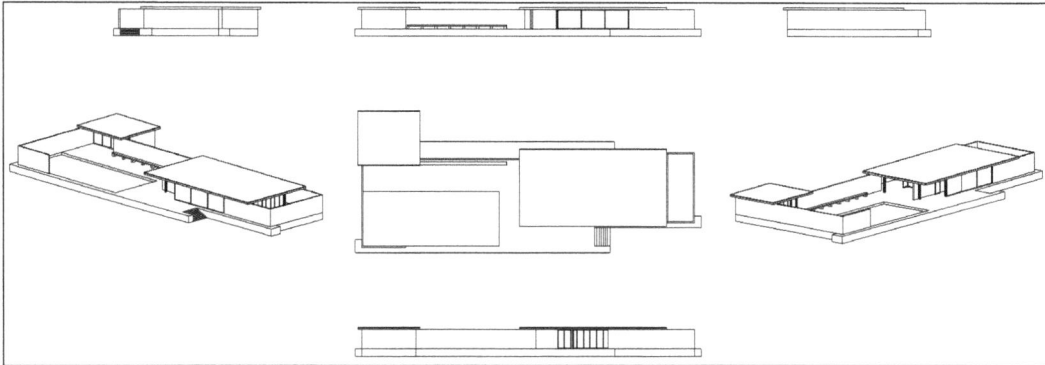

Exercise 9.3

We continue the model from *Exercise 9.2*, now creating some flattened and projected views.

1. Open the drawing A3D_09_3.DWG.

2. Projected views should be on a proper layer, so create and activate a layer called **2DVIEWS**.

3. With the **ViewCube** tool, the **Front** view is specified.

4. Applying the FLATSHOT command, on the dialog box confirm **Insert as new block**, uncheck **Obscured Lines Show**, and check **Include tangential edges**. Upon pressing **Create**, specify an insertion point not far from the model for the new block and confirm scales and rotation.

5. Specify the **Top** view and repeat the FLATSHOT command with the same parameters. Place the new block on the same plane and near the previous block.

6. Repeat the command with the **Left** and **Right** views and a view from the third quadrant. Arranging the blocks, you should have a display similar to the next diagram (first angle projection, also known as European projection).

7. Save the drawing with the name A3D_09_3final.DWG:

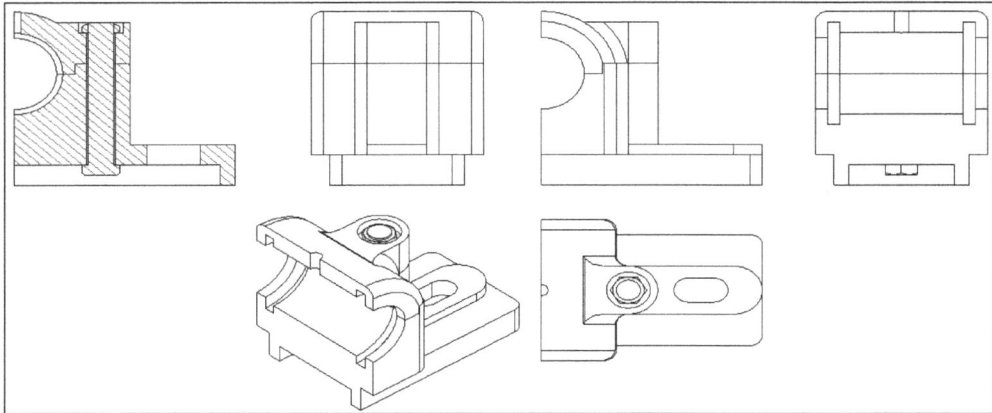

Associative views

Some of the most important innovations in AutoCAD Versions 2012 and 2013 are associative views. These associative views are new AutoCAD objects known as drawing views and represent dynamic 2D projections, sections, or details from the 3D model. If the 3D model changes, these associative views update accordingly.

Access and automatic layers creation

The majority of the commands and functions related to associative views are not available on toolbars or the menu bar, but can be found on the ribbon, **Layout** tab, and the following panels: **Create View**, **Modify View**, **Update**, and **Styles and Standards**. So, for users who rather work with a classic workspace, they can change to the **3D Modeling** workspace for these commands and, then, come back to the **Classic** workspace:

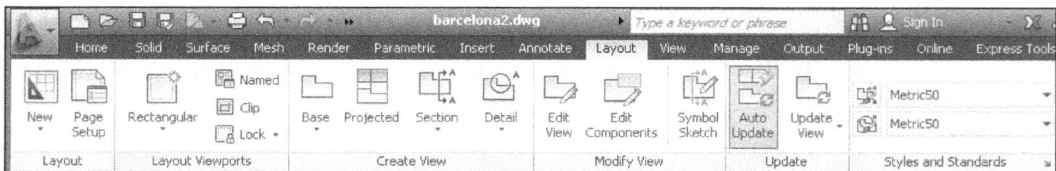

When creating these associative views, AutoCAD creates and applies the following layers, in Version 2012, all with color 7 (white):

- **Hidden**: This layer has the **HIDDEN2** linetype and lineweight **0.35**
- **Hidden Narrow**: This layer has the **HIDDEN2** linetype and lineweight **0.25**
- **Visible**: This layer has the **CONTINUOUS** linetype and lineweight **0.50**
- **Visible Narrow**: This layer has the **CONTINUOUS** linetype and lineweight **0.35**

In version 2013, there are six layers, all with color 7:

- **MD_Annotation**: This layer has the **CONTINUOUS** linetype and lineweight **0.25**
- **MD_Hatching**: This layer has the **CONTINUOUS** linetype and lineweight **0.25**
- **MD_Hidden**: This layer has the **HIDDEN2** linetype and lineweight **0.35**
- **MD_Hidden Narrow**: This layer has the **HIDDEN2** linetype and lineweight **0.25**
- **MD_Visible**: This layer has the **CONTINUOUS** linetype and lineweight **0.50**
- **MD_Visible Narrow**: This layer has the **CONTINUOUS** linetype and lineweight **0.35**

Base and projection views

Following are the main commands for creating and editing associative views.

The VIEWBASE command

The VIEWBASE command (no alias, **Create View** panel on the ribbon) creates a base associative view in a layout, from solids and surfaces. When dealing with associative views, this must be the first command to execute. After a base view is specified, the command allows you to create projected views. With Version 2013, the command can be used in model space, with the previous version only in layouts.

The command starts by prompting whether we want the existing model (**Model space**) or a model from an Autodesk Inventor file (**File**):

```
Command: VIEWBASE
Specify model source [Model space/File] <Model space>: Enter
```

Next, we can select the objects that will be processed by the associative views, or press the *Enter* key to select the entire model:

```
Select objects or [Entire model] <Entire model>: Enter
```

If using the command in model space, the command prompts for the name of a new or existing layout:

```
Enter new or existing layout name to make current or [?] <Test>: Name
```

Then, we access the chosen layout and the command displays the default values for the base view representation:

```
Type = Base and Projected  Hidden Lines = Visible and hidden lines
Scale = 1:1
```

The location of the base view is prompted. By default, it applies the model **Front View** and is represented with a shaded visual style:

```
Specify location of base view or [Type/sElect/Orientation/Hidden
lines/Scale/Visibility] <Type>: Point
```

We may change the view options or press the *Enter* key to continue:

```
Select option [sElect/Orientation/Hidden lines/Scale/Visibility/Move/
eXit] <eXit>: Enter
```

The command prompts for projected views. The cursor position indicates which view will be obtained. We define several points as follows:

```
Specify location of projected view or <eXit>: Point
Specify location of projected view or [Undo/eXit] <eXit>: Point
Specify location of projected view or [Undo/eXit] <eXit>: Point
Specify location of projected view or [Undo/eXit] <eXit>: Point
Specify location of projected view or [Undo/eXit] <eXit>: Point
```

The output of these commands is shown in the following diagram:

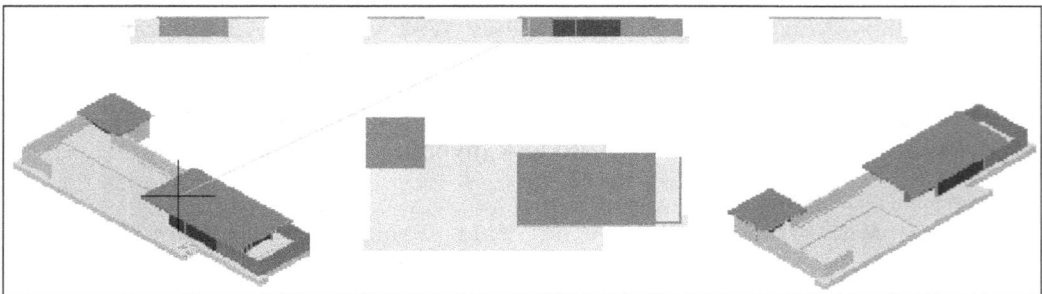

We press the *Enter* key to finish. All views are then calculated according to the default options as follows:

```
Specify location of projected view or [Undo/eXit] <eXit>: Enter
```

The output is displayed in the following diagram:

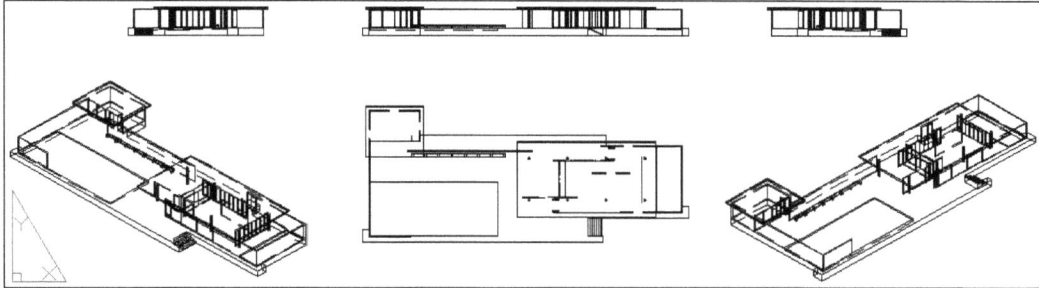

This command has the following options:

- **File**: This option allows selecting an IAM, IPT, or IPN Inventor file. The model is not imported, it is only used to create the associative views.

- **Type**: This option specifies whether the VIEWBASE command is only used to create the base view, or if it continues to create the projected views.

- **sElect**: This option allows you to select the objects that will be processed by the command. This option does not exist in version 2012.

- **Orientation**: This option allows you to choose the orientation of the base view, between the current one, one of the six orthographic views, or one of the four isometric views.

- **Hidden lines**: This option allows you to choose the view representation between **Visible lines**, **Visible and hidden lines**, **Shaded with visible lines**, and **Shaded with visible and hidden lines**:

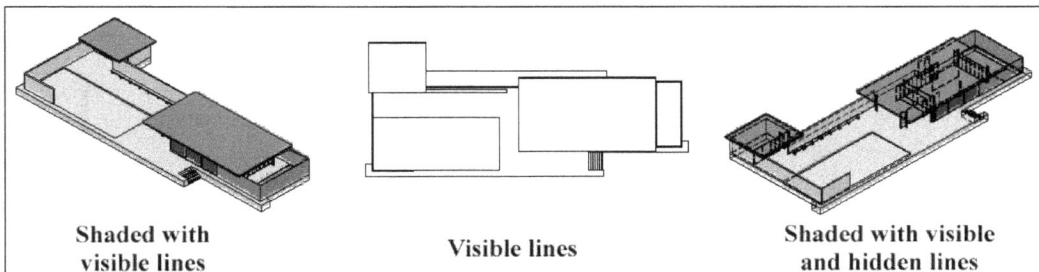

Shaded with visible lines   Visible lines   Shaded with visible and hidden lines

- **Scale**: This option specifies the view scale. All projected views from this will get the same scale.
- **Visibility**: This option specifies options for including interference edges and tangent edges. Other options are available for Inventor models.
- **Move**: We may move the base view without ending the command.

> The default scale used by the command is based on the paper dimension. To a more efficient application of associative views, it is better to first configure the layout and apply the command only after that.
>
> The quality of shaded associative views is not very good, therefore, should be avoided.

The VIEWPROJ command

The VIEWPROJ command (no alias, **Create View** panel on the ribbon) creates a projected associative view in a layout. Another process for applying the command is using the shortcut menu. It prompts for an existing view and the location of the projected views:

```
Command: VIEWPROJ
Select parent view: Selection
Specify location of projected view or <eXit>: Point
```

When ending the command, the view is processed:

```
Specify location of projected view or [Undo/eXit] <eXit>: Enter
1 projected view(s) created successfully.
```

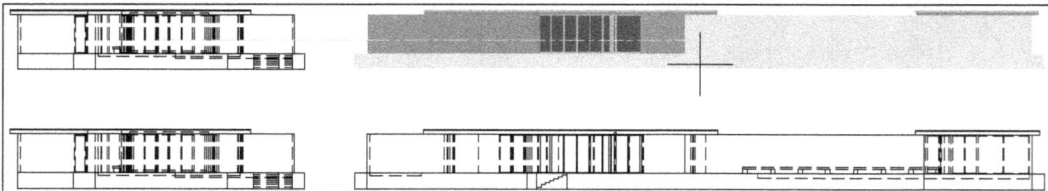

> If we need a Model **Back View**, we may apply this command, selecting the left or right views as the base view.

The VIEWEDIT command

The VIEWEDIT command (no alias, **Modify View** panel on the ribbon) allows you to edit an associative view. The command prompts for the selection of the view to edit:

```
Command: VIEWEDIT
Select view: Selection
```

Then, we choose what to change:

```
Select option [sElect/Hidden lines/Scale/Visibility/eXit] <eXit>:
Option
```

Options for this command, when using AutoCAD objects:

- **sElect**: This command allows you to select objects to add to views or remove objects from views. It can only be applied to the base view. This option does not exist on Version 2012.

- **Hidden lines**: This command allows you to choose the selected view representation between **Visible lines**, **Visible and hidden lines**, **Shaded with visible lines**, and **Shaded with visible and hidden lines**.

- **Scale**: This command specifies the view scale. If the selected view is the base view, all projected views get the same scale. If the selected view is a projected one, the scale modification is only for that view.

- **Visibility**: This command specifies options for including interference edges and tangent edges.

The view scale for the base view can also be modified with the PROPERTIES or QUICKPROPERTIES commands. An easier method is to select the view and apply the small triangular grip to get a list of all available scales.

When using a ribbon interface, the VIEWEDIT command also displays a contextual grip, with all important options for the selected view:

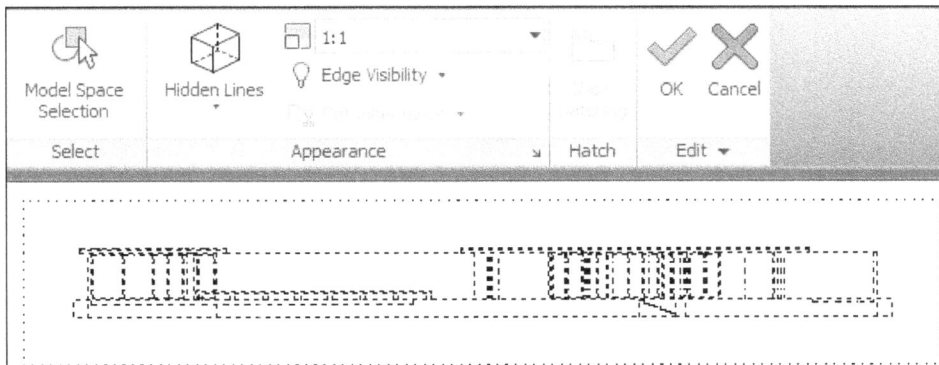

The VIEWSTD command

There are some standards to create projected views. The `VIEWSTD` command (no alias, **Styles and Standards** panel on the ribbon) displays a dialog box to specify these standards. It includes the following areas:

- **Projection type**: We choose between **First angle** projection (also known as European projection) or **Third angle** projection (also known as American projection).

- **Thread style**: Only for Inventor models, it allows to specify the thread representation.

- **Shading/Preview**: **Shaded view quality** allows to control the view quality in dots per inch when choosing **Shaded** options. **Preview type** allows specifying the representation of the preview view, between **Shaded** and **Bounding box**:

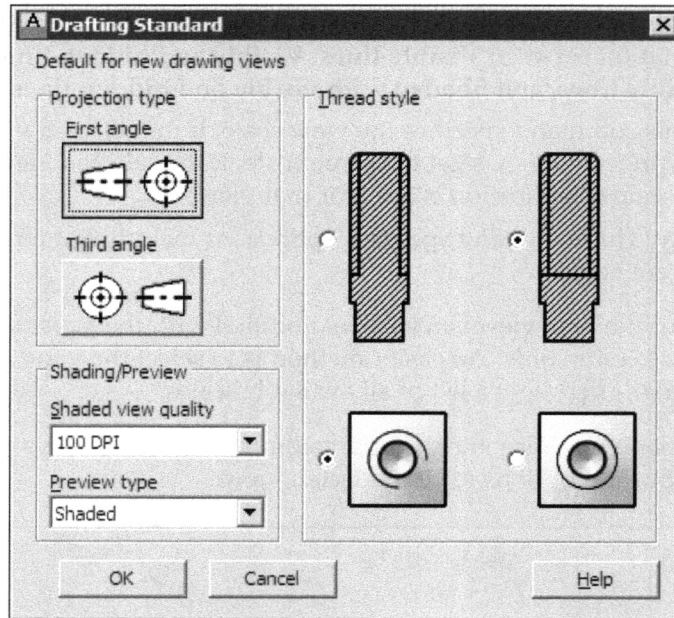

The VIEWUPDATE command

The VIEWUPDATE command (no alias, **Update** panel on the ribbon) allows you to update associative views. The command prompts for the view selection to update and repeat the prompt until we cancel the command:

```
Command: VIEWUPDATE
Select view to update: Selection
```

> The VIEWUPDATEAUTO variable, new on Version 2013, controls automatic updates. With value 1, view updates turns on automatically whenever the model is modified.

Section and detail views

Version 2013 adds associative section and detail views, including proper view styles.

The VIEWSECTION command

The VIEWSECTION command (no alias, **Create View** panel on the ribbon) allows you to create sections of solids and surfaces in AutoCAD or Inventor models. Another process for applying the command is using the shortcut menu. The command starts by prompting the view selection that will define the section:

```
Command: VIEWSECTION
Select parent view: Selection
Hidden Lines = Visible lines  Scale = 1:1 (From parent)
```

We specify two or more points that will define the section plane. The default section view style controls the reference letter and view label:

```
Specify start point or [Type/Hidden lines/Scale/Visibility/Annotation/
hatCh] <Type>: Point
Specify next point or [Undo]: Point
Specify next point or [Undo/Done] <Done>: Enter
```

We pick the location of the view, the new view is displayed and we may change some options:

```
Specify location of section view or:
Select option [Hidden lines/Scale/Visibility/Projection/Depth/
Annotation/ hatCh/Move/eXit] <eXit>: Point
```

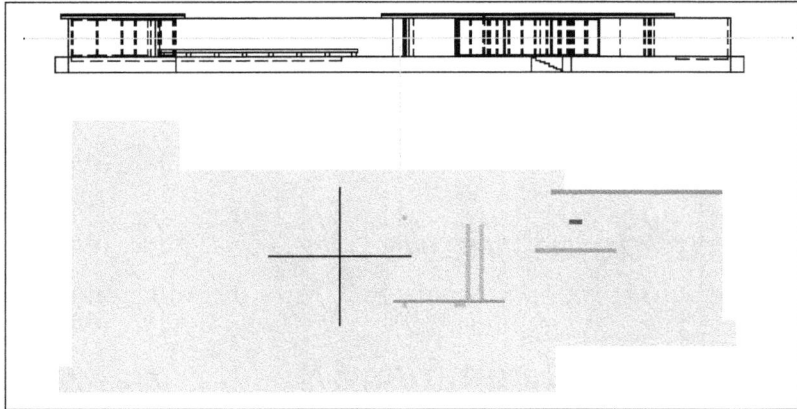

Pressing the *Enter* key, the view is processed:

```
Section view created successfully.
```

The following image is then displayed:

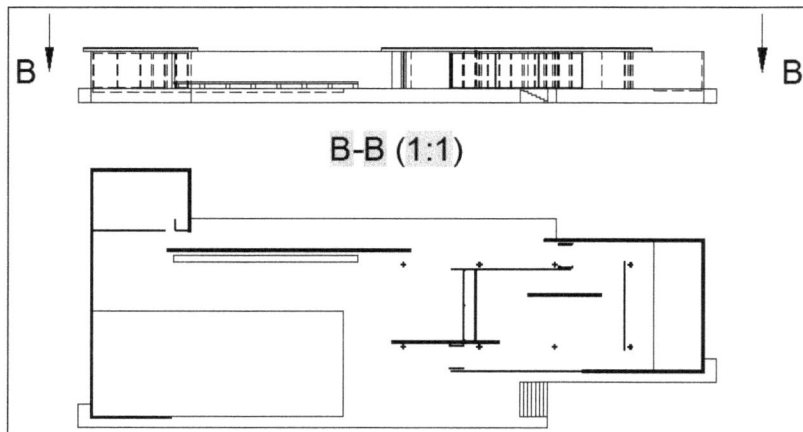

Most options are similar to the VIEWBASE or VIEWEDIT commands, but some are specific to this command:

- **Type**: This option specifies the section type. **Full** defines a complete section by two points; **Half** defines only half section, the remaining is like a projected view; **Offset** defines a section with several planes; **Align** defines an aligned section; and **Object** prompts for an object selection that will define the section plane.

- **Annotation**: This option allows you to modify the default letter that identifies the section if the label is displayed.

- **hatCh**: This option specifies whether the section view includes hatches or not.

- **Projection**: This option specifies the type of projection, if it is orthographic (by default) or perpendicular to the section plane.

- **Depth**: This option specifies the depth in which visible edges are displayed.

> By default, the section view is placed perpendicular to the definition view. But, by pressing and releasing the *Shift* key, it can be placed anywhere.

The VIEWSECTIONSTYLE command

When dealing with associative section views, several aspects are controlled by section view styles, similar to other AutoCAD styles. The VIEWSECTIONSTYLE command (no alias, **Styles and Standards** panel on the ribbon) allows you to create and managing section view styles. The command displays the **Section View Style Manager** dialog box with a preview image, where you can activate, create, modify, or delete section view styles.

When creating or modifying section view styles, a new dialog box is displayed with four tabs:

- **Identifier and Arrows**: This tab formats the letter identifying the section, arrows, and relative positions

- **Cutting Plane**: This tab formats lines defining the section plane, end, and bending lines, and turns them normally thicker

- **View Label**: This tab formats the label placed near the section view
- **Hatch**: This tab formats the hatch corresponding to the cut part

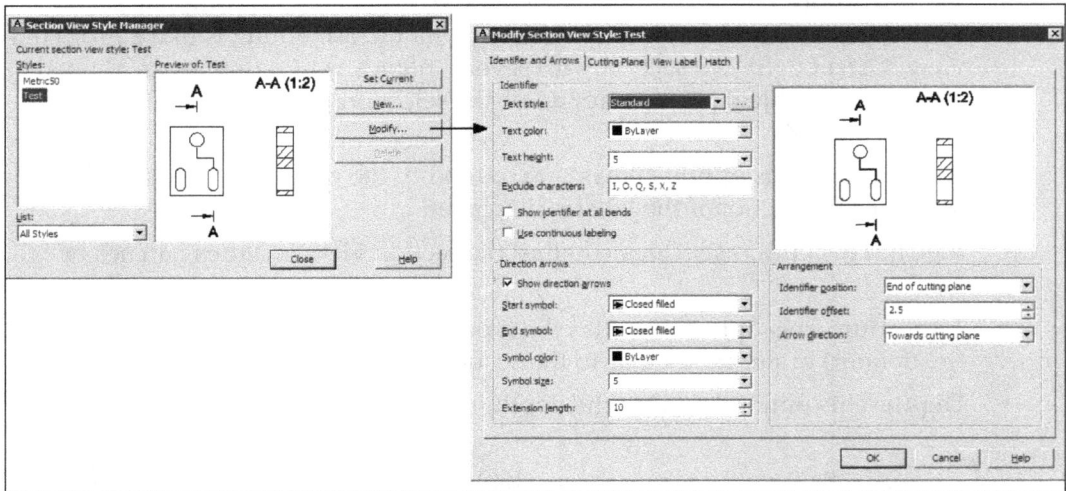

> Unlike other styles, the section styles list on the ribbon only allows you to specify the default section style for new sections. It does not allow you to change existing sections.
>
> Section view styles can be imported from other drawings with the **Design Center**, using *Ctrl + 2*.
>
> Default styles **Metric50** and **Imperial24** cannot be renamed or deleted.

The VIEWCOMPONENT command

In some areas, cylindrical elements, such as shafts and bolts, should not be sectioned. The VIEWCOMPONENT command (no alias, **Modify View** panel on the ribbon) allows you to control which elements should not be sectioned. The command prompts for the selection of objects, which must be on a section view:

```
Command: VIEWCOMPONENT
Select component: Selection
```

Then, we select one of three options:

```
Select section participation [None/Section/sLice] <Section>: Option
```

The options are:

- **None**: Using this option the component is not sectioned
- **Section**: Using this option the component is again sectioned, if it was previously removed
- **sLice**: Using this option the component is sectioned and background edges are not represented

The VIEWDETAIL command

The VIEWDETAIL command (no alias, **Create View** panel on the ribbon) allows you to create details of AutoCAD or Inventor models. Another process for applying the command is using the shortcut menu. The detail can be circular or rectangular and is created with a default detail view style. The command starts by prompting the view (base or projected) where we want the detail:

```
Command: VIEWDETAIL
Select parent view: Selection
```

It displays default options and prompts for the center of the detail:

```
Boundary = Circular  Model edge = Smooth  Scale = 2:1
Specify center point or [Hidden lines/Scale/Visibility/Boundary/model
Edge/Annotation] <Boundary>: Point
```

Then, the dimension of the detail is requested:

```
Specify size of boundary or [Rectangular/Undo]: Point or value
```

We must now specify the location of the detail view:

```
Specify location of detail view: Point
```

Pressing the *Enter* key calculates the detail as follows:

```
Select option [Hidden lines/Scale/Visibility/Boundary/model Edge/
Annotation/Move/eXit] <eXit>: Enter
```

Again, most options are common to the creation of other views, with the following exceptions:

- **Boundary**: This option allows you to choose between a circular or a rectangular detail
- **Annotation**: This option allows you to modify the default letter that identifies the detail and if the label is displayed
- **model Edge**: This option specifies whether the section view includes hatches or not

This is displayed in the following diagram:

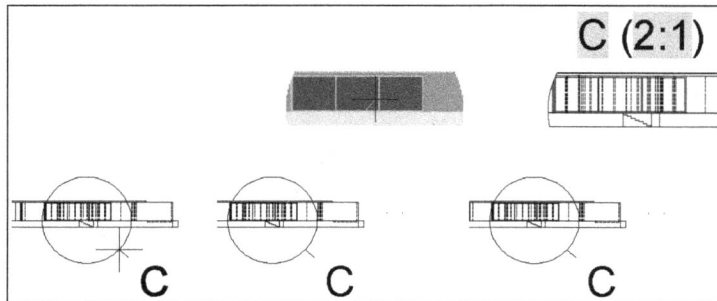

The VIEWDETAILSTYLE command

Like section views, we have detail view styles. The VIEWDETAILSTYLE command (no alias, **Styles and Standards** panel on the ribbon) allows you to create and manage detail view styles. The command displays the **Detail View Style Manager** dialog box with a preview image, where we can activate, create, modify, or delete detail view styles.

When creating or modifying detail view styles, another dialog box is displayed, with three tabs:

- **Identifier**: This tab formats the letter identifying the detail and its placement
- **Detail Boundary**: This tab formats the boundary line, the model edge, and the connection line
- **View Label**: This tab formats the label placed near the detail view

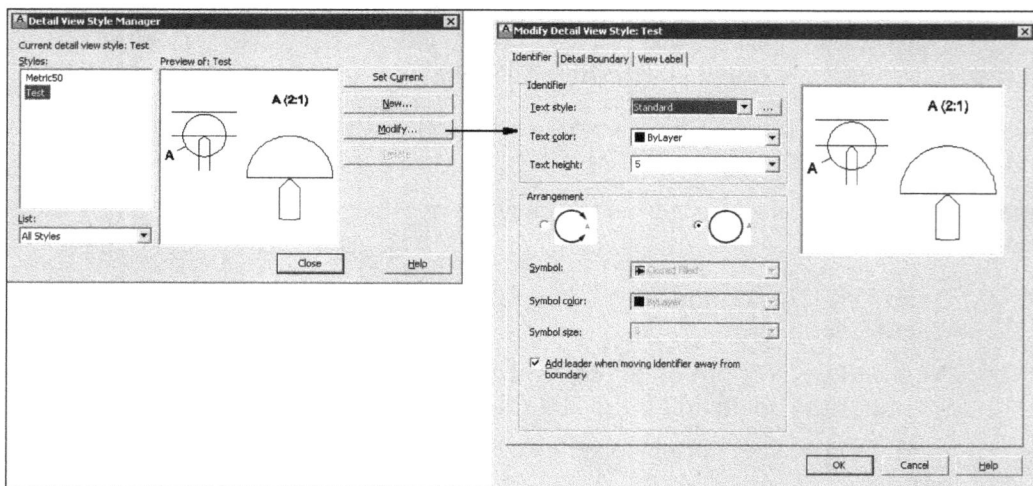

> Unlike other styles, the detail styles listed on the ribbon only allow you to specify the default detail style for new details. It does not allow you to change existing details.
>
> Detail view styles can be imported from other drawings with the **Design Center**, *Ctrl + 2*.
>
> Default styles **Metric50** and **Imperial24** cannot be renamed or deleted.

Other commands

Related to associative views, there are some less important commands, listed next:

- EXPORTLAYOUT: This command has been available since Version 2009, it is not exclusive to associative views, but can be very useful if we want to make nonassociative changes. The command exports the active layout to the model space of a new drawing.

- VIEWSYMBOLSKETCH: This command allows defining a connection between the section lines or boundary detail lines to the model geometry. If the model is modified, these lines will adjust accordingly.

- VIEWSKETCHCLOSE: This command ends the connection defined with the previous command.

Exercise 9.4

In the model coming from the previous exercises, we will create several associative views:

1. Open the drawing A3D_09_4.DWG.

2. As associative views are to be used on layouts, first we must configure a layout to be used. Use **Layout1** and rename it to **AssocViews**. If a viewport is displayed, delete it. With PAGESETUP, configure the following: **Printer/plotter** as **DWG to PDF** and **Paper size** as **ISO full bleed A3 (420.00 x 297.00 MM)**. As this assembly is in mm, there is no need to change scale.

3. Apply the VIEWBASE command and press the *Enter* key to specify **Model space**. Pick on the top part of the layout and press the *Enter* key to accept options. By default, the command is followed by projected views, so specify top, left, right, and one perspective view.

4. Adjust views positions. To decrease line-type scale, assign 1 to LTSCALE:

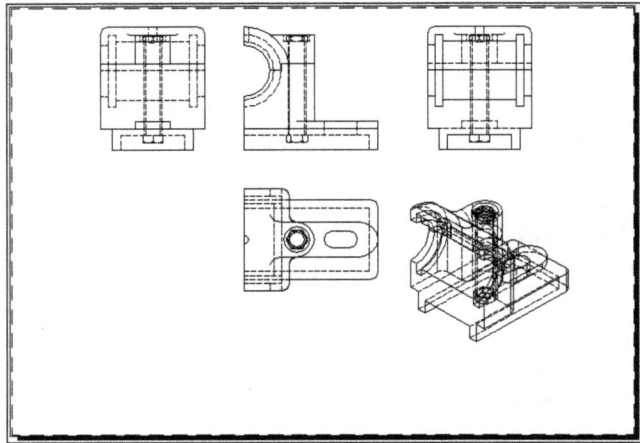

5. The perspective view is a bit confusing, so we select this and with the shortcut menu (or ribbon) apply **Edit View** (VIEWEDIT command). Under the **Hidden** lines option, choose **Visible lines**.

6. We are going to create a section view by applying the shortcut menu to the top view, choosing **Section View** (VIEWSECTION command) inside **Create Views**. Define the section line in the middle by two points to the left and right of midpoints. When prompting for the third point, press the *Enter* key. Next, we have to define the section position and, as we don't want it to be orthogonal to the top view, press and release the *Shift* key to place it to the left of the top view. Move the view label closer to the view:

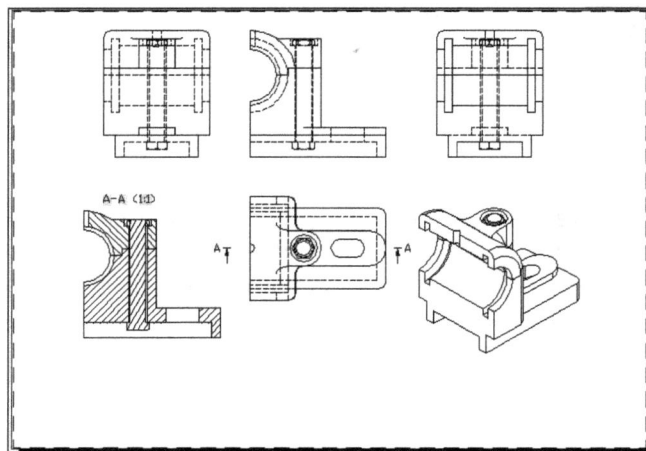

7. Normally, we don't section nuts and bolts, so we need to remove hatches on them. Apply the `VIEWCOMPONENT` command (for instance, with **Edit Components** on the ribbon). Select both elements on the section view and apply the **None** option.

8. Finally, we are going to apply a detail. Select the top view and, with the shortcut menu, choose **Detail View** inside **Create View**. Select the center of the nut and define a radius of about `18mm`. Place the view below the Top View and confirm options with the *Enter* key. Adjust the label position and that's all:

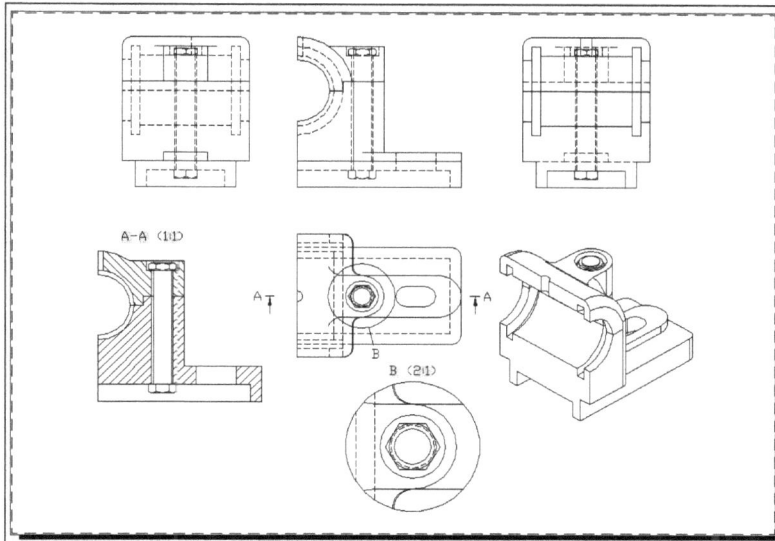

9. Save the drawing with the name `A3D_09_4final.DWG`.

Summary

The subject of this chapter is obtaining 2D drawings from 3D models. After the creation of a 3D model, 2D projections, sections, and details are common requirements.

For the creation and configuration of layouts, namely shortcut menus, the LAYOUT and PAGESETUP commands, namely shortcut menus, are suitable. To create viewports, we have the MVIEW command and several tools to adjust their properties and scales. To control colors and lineweights, we apply plot styles and also may control layer properties per viewport. The PREVIEW command allows verifying the print, and PLOT finally prints the layout.

To create projection views and sections (not associative), we have the following commands: SECTIONPLANE to obtain sections; SECTION also to obtain sections (simpler but less flexible); and FLATSHOT, to obtain projection views.

To end the chapter, we present associative views, which are linked to the 3D model. VIEWBASE command allows you to create a base and projected views; VIEWPROJECT creates projected views from an existing view; VIEWEDIT allows you to edit existing views; VIEWSTD specifies some drawing standards, and VIEWUPDATE allows you to update views, if not automatic. To create section views we have VIEWSECTION and VIEWSECTIONSTYLE to create and manage styles and VIEWCOMPONENT to remove parts from sectioning. Details can easily be obtained with the VIEWDETAIL command and respective styles with the VIEWDETAILSTYLE command.

In the next chapter we will see how to obtain renders and realistic images and how to set up illumination in our scene.

10
Rendering and Illumination

After creating a 3D model, it is time to present it as a virtual prototype: how it will look when built or fabricated. In this chapter we will present the rendering process and all related commands, as well as simulating natural and artificial lighting.

The topics covered in this chapter are as follows:

- Understanding rendering
- How to create photorealistic images
- How to create test renders
- Identify generic and photometric lighting
- How to illuminate a scene
- How to apply lighting from the Sun and the sky
- How to apply point lights, spot lights, and other lights
- How to manage and adjust lights

Rendering concepts and commands

We present all rendering concepts and commands to obtain photorealistic images from our 3D model.

General concepts

A fundamental output in all 3D models is the creation of photorealistic images, which represent how the model will look after building or fabricating it. A render is, thus, an image, which is defined in horizontal and vertical pixels obtained from a 3D model, where AutoCAD calculates the visible faces and all lighting in the scene and eventually includes indirect lighting, applied materials, and some effects such as background.

After creating the 3D scene with all commands and methods already seen, a typical rendering workflow is as follows:

1. Apply the camera's simulating user location, point to look at (target), and perspective defining what we want to show.

2. Apply the lighting conditions. For exterior scenes or interior scenes where outside light is important, light due to the Sun and the sky may be considered. For artificial lights, we apply point lights, spot lights, or web lights that simulate real conditions.

3. Apply realistic materials. AutoCAD comes with a large library of materials that can be applied right away, or adjusted easily. However, it is quite easy to create and configure materials, from photos, images, or procedural maps. Materials and textures are presented in the next chapter.

4. Apply environment effects, such as background, fog, or depth cue.

5. Apply render commands.

AutoCAD uses the mental ray rendering technology, licensed to a German company called Mental Images, later bought by NVIDIA. This powerful renderer supports physically correct simulation of lighting, including indirect lighting. The next image includes an example.

> Rendering is an iterative process. Only with renders can we adjust illumination and materials in order to get the desired realism. After lighting or material changes, it is common to render the scene or a part of it. These intermediate renders don't need full quality.
>
> Render time depends on the complexity of the model, number of lights, indirect lighting, applied materials (reflections and refractions take longer), render quality, and render resolution.

Access to commands

All major commands related to rendering, lighting, and materials are in the
Render panel on the ribbon, which can be found on the menu bar by going to
View | Render and on the **Render** toolbar.

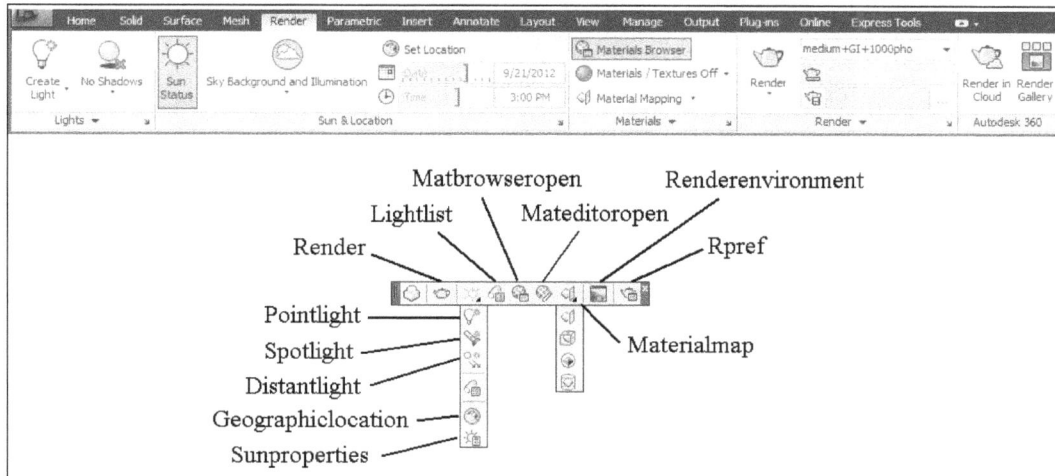

Rendering commands

In the following sections we shall see all commands that are related to rendering
calculations, including presets and saving the image to a file.

The RENDER command

The RENDER command (alias RR) creates a realistic image according to the 3D model,
lighting, materials, effects, and all default render parameters. The render parameters,
such as image resolution, are defined with the RPREF command.

When applying this command, the render starts immediately, directly on the
drawing area or on a special **Render** window, depending on the default parameters.

The **Render** window (displayed in the following screenshot) has three resizable areas and a menu bar:

- **Image**: This area displays the image being calculated. The calculation is made in small squares, typically by the mental ray rendering method. Below the image are two horizontal bars that display the render progress meter: the one on the top indicates the progress of the current render phase and the one on the bottom shows the complete render.

- **History**: This area, below the image, includes all the rendered images in the drawing session, allowing for easy comparisons. Columns include the output size in pixels, view name, render time, and render preset. Tools (presented in the following topic) are provided to save images.

- **Image Information**: This area displays all the default settings used for the selected image, including some statistical information.

Menus and saving images

On the top menu bar, under **File**, we can save the image or save a copy of it under another name; **View** allows the display of the status bar or the statistics area; **Tools** allows zooming in to or out of the image (the mouse wheel also allows for this).

On the bottom area, by right-clicking on an image listed, we access the shortcut menu with the following options:

- **Render Again**: This replaces the selected render with a render from the current view and settings.

- **Save...**: This saves the selected drawing in the JPG, PNG, TIF, TGA, or BMP format. After name and location specifications, a box is displayed with options related to the chosen format, namely the number of colors and the resolution.

- **Save Copy...**: This creates another image from the same render. The name must be different, even if we choose another format.

- **Make Render Settings Current**: This allows us to specify the settings from the selected render for the next renders.

- **Remove From the List**: This removes the selected render from the list, but if the render was already saved, the corresponding image file is not deleted.

- **Delete Output File**: If the file was saved, this option allows us to delete the image file.

The shortcut menu can be seen in the following screenshot:

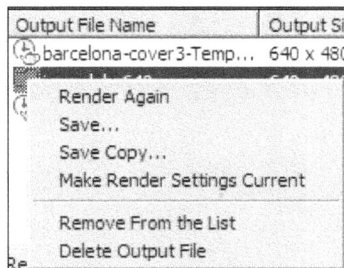

The RENDERWIN command

The RENDERWIN command (alias RW) displays the **Render** window, allowing us to view the processed renders in the current drawing session. If we apply this command, we don't need to create or repeat the last render; just the render window with the render history is displayed. The command asks nothing.

The RENDERCROP command

The RENDERCROP command (alias RC) allows us to create a render of a part of the current view directly on the viewport and not on the **Render** window. It just prompts for two opposite corners and starts the render immediately.

```
Command: RENDERCROP
Pick crop window to render: Point
please enter the second point: Point
```

> This command is great for testing materials and parts of a scene without needing to render the complete view.

The following screenshot shows the result of the RENDERCROP command:

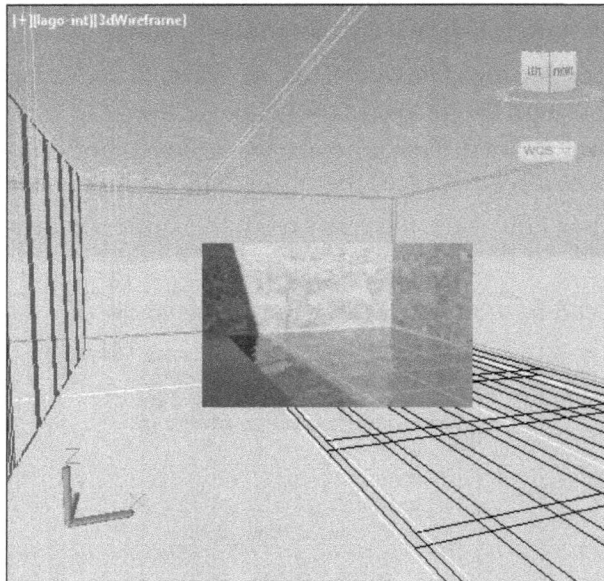

The SAVEIMG command

The SAVEIMG command (no alias) saves the current render to a file, displaying the file dialog box, and then according to the chosen format, the settings dialog box.

> When rendering to the viewport, this command is useful if we want to save the resulting render.

The RPREF command

The RPREF command (alias RPR) is the command specifying the default settings for rendering calculations with the RENDER or RENDERCROP command. It displays the **Advanced Render Settings** palette, which means that it can stay visible with multiple parameters.

On the top we have a list of all the saved render presets and the **Render** button. In all the scenes we have five predefined rendering presets ranging from low to very high quality. The higher the quality, the longer the rendering. The **Draft** and **Low** preset qualities are very bad, **Medium** is acceptable, **High** is very good, and **Presentation** is the best. Additionally, we can define new preset settings.

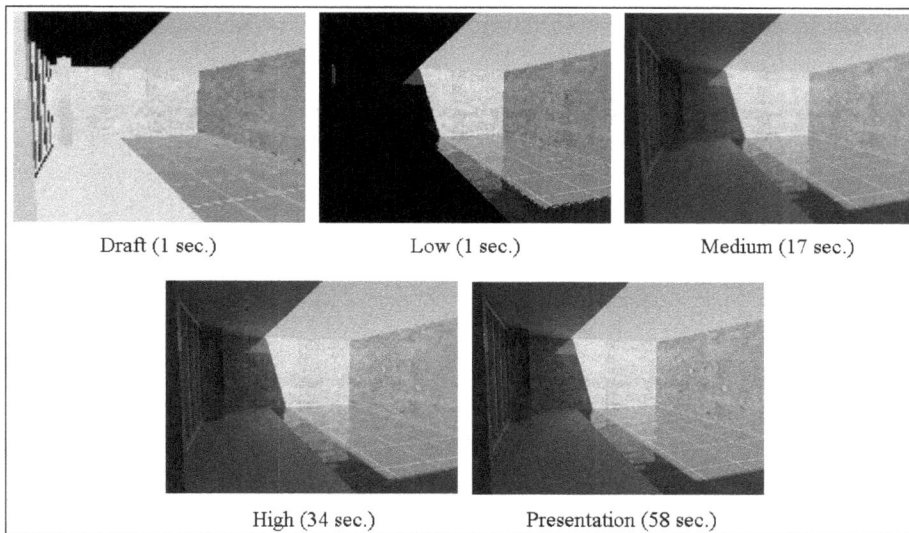

Draft (1 sec.)   Low (1 sec.)   Medium (17 sec.)

High (34 sec.)   Presentation (58 sec.)

The **General** panel, as displayed in the following image, contains the following areas:

- **Render Context**: This contains the most important parameters. The small icon allows us to save the file when rendering. In **Procedure**, we choose to render the current view, crop the view, or only render the selected objects. **Destination** indicates if the render is processed on the render window or on the viewport. When saving the file, **Output file name** allows us to specify the name and the folder. **Output size** specifies the render resolution, from predefined sizes or as defined by us; the maximum size is 4000 x 4000 pixels. **Exposure Type**, which simulates camera's exposure, can be automatic or logarithmic. **Physical Scale** indicates the physical scale applied to lights.

- **Materials**: **Apply materials** allows us to render with the applied materials; if set to **Off**, it will render with GLOBAL materials that use the object's color. **Texture filtering** controls the antialiasing applied to textures. **Force 2-sided** applies materials on both sides of faces.

- **Sampling**: This area controls antialiasing effects, thus avoiding jagged edges. We define the minimum and maximum number of samples that will define each pixel, the sample mechanism, the number of pixels in width and height of the interference area, and the influence of colors in antialiasing.

- **Shadows**: The lamp icon turns the shadows' calculation on or off. **Mode** defines a transparent shadow's calculation. **Shadow map** allows for soft-edged shadows. **Sampling Multiplier** limits sampling for area lights.

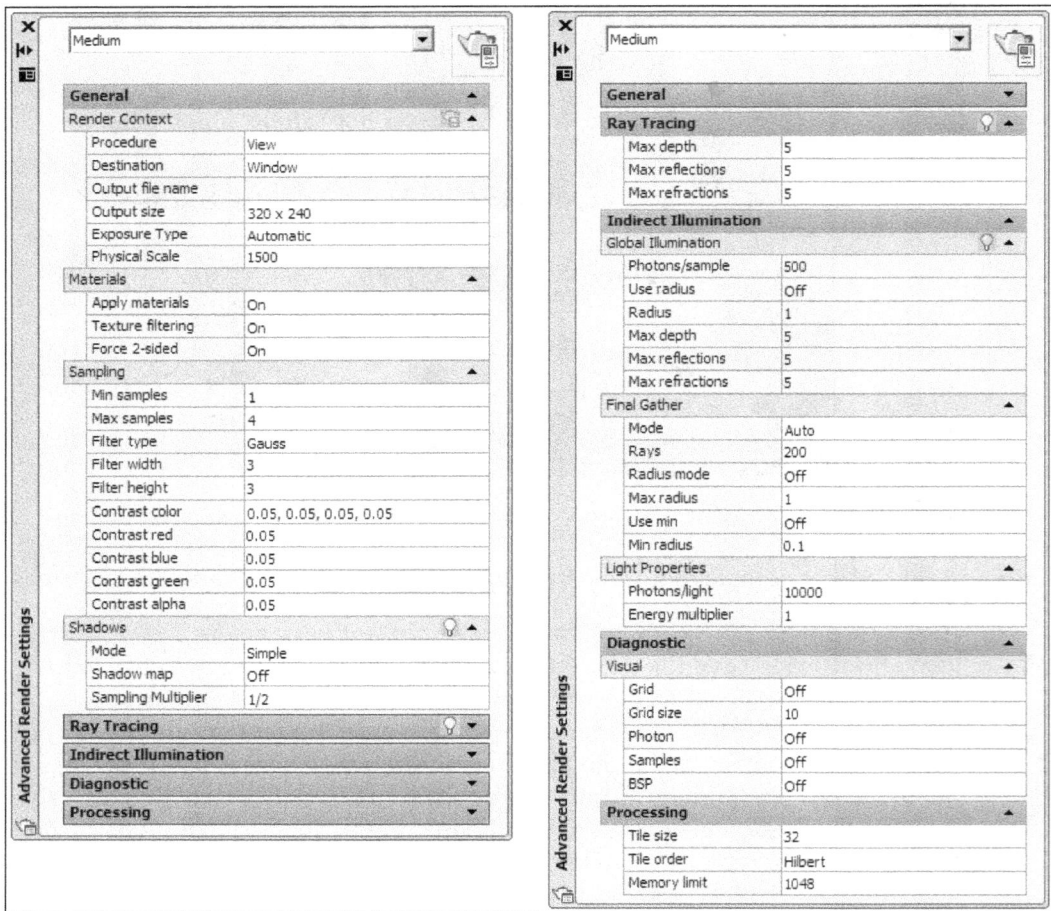

General	
Render Context	
Procedure	View
Destination	Window
Output file name	
Output size	320 x 240
Exposure Type	Automatic
Physical Scale	1500
Materials	
Apply materials	On
Texture filtering	On
Force 2-sided	On
Sampling	
Min samples	1
Max samples	4
Filter type	Gauss
Filter width	3
Filter height	3
Contrast color	0.05, 0.05, 0.05, 0.05
Contrast red	0.05
Contrast blue	0.05
Contrast green	0.05
Contrast alpha	0.05
Shadows	
Mode	Simple
Shadow map	Off
Sampling Multiplier	1/2
Ray Tracing	
Indirect Illumination	
Diagnostic	
Processing	

General	
Ray Tracing	
Max depth	5
Max reflections	5
Max refractions	5
Indirect Illumination	
Global Illumination	
Photons/sample	500
Use radius	Off
Radius	1
Max depth	5
Max reflections	5
Max refractions	5
Final Gather	
Mode	Auto
Rays	200
Radius mode	Off
Max radius	1
Use min	Off
Min radius	0.1
Light Properties	
Photons/light	10000
Energy multiplier	1
Diagnostic	
Visual	
Grid	Off
Grid size	10
Photon	Off
Samples	Off
BSP	Off
Processing	
Tile size	32
Tile order	Hilbert
Memory limit	1048

The **Ray Tracing** panel allows us to control the maximum number of reflections and refractions for rays used in render calculation, with **Max depth** being the sum of reflections and refractions.

The **Indirect Illumination** panel contains the following areas:

- **Global Illumination**: With the lamp turned on, indirect lighting is considered, that is, light reflects on objects. With this turned off, areas without direct lighting will render as black. **Photons/sample** specifies how many photons per sample are considered. **Use radius**, if on, specifies a radius for samples. The remaining parameters define the photons' number of reflections and refractions; larger values will cause brighter indirect illumination, but will need more time for rendering.

- **Final Gather**: This is another mechanism for processing indirect lighting, where rays are shot from the user view (or camera). It is normally faster than global illumination. We control the number of rays and also if a sample radius will be used, and its maximum and minimum values.

- **Light Properties**: We can specify the number of photons per light and an energy multiplier.

The **Diagnostic** panel includes some advanced tools for analyzing the render mechanism.

Finally, the **Processing** panel specifies the size, method, and memory limit for the small squares (buckets) used in render calculations.

> For most users, it is enough to apply the **Render Context** parameters and choose a preset render quality. While improving the scene, the **Medium** preset is okay. For final renders, the **High** or **Presentation** preset should be used, with the latter advised when the materials' textures don't look good enough.

The RENDERPRESETS command

The RENDERPRESETS command (alias RP) allows us to define preset render qualities. Besides the five presets available, we may define any number of new presets. The command displays a dialog box with three areas. On the left is the list of all the available presets. All parameters that are identical to the RPREF parameters of the selected preset are displayed on the center and can be adjusted. On the right are the buttons to activate a preset, to create a new from the selected one, and to delete a custom preset.

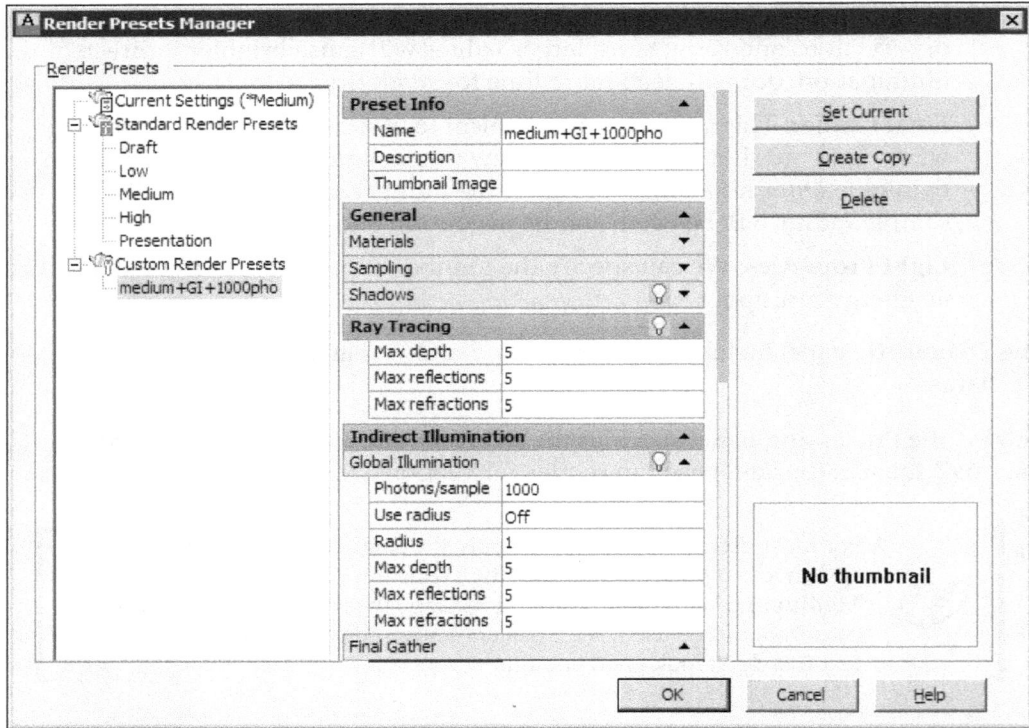

Exercise 10.1

We are going to try several render configurations with a simple drawing:

1. We open the drawing A3D_10_01.DWG. This drawing contains a thick glass bowl on a table and some shiny balls.

2. Without changing anything, we make a first render by typing in RR.

3. Applying the RPREF command, we try the **Draft** and **Low** modes. Instead of applying the RENDER command, we may press the **Render** button to the right of the list.

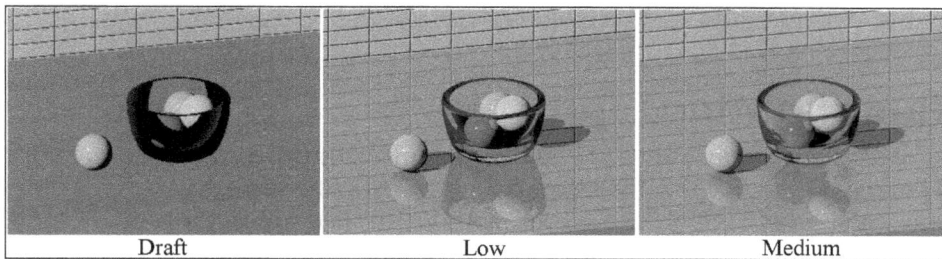

| Draft | Low | Medium |

4. Now, let's try the **High** and **Presentation** modes. On the render window, we can compare all these renders (including time), and see improvements, especially on the wall and bowl.

5. Maintaining the render preset in **Presentation**, in the **Final Gather** area, we change **Mode** to **On** and assign 5000 rays. The render improves a little bit, but the time taken increases by more than seven times.

6. We set the **Final gather mode** option to **auto** and decrease the number of rays to 200. Now we turn on **Global Illumination** and render again. We notice that darker areas get lighter because there is some light reflected off the wall. Render time decreases a lot as compared to **Final Gather** with 5000 rays.

7. We save the drawing with the name A3D_10_01final.DWG.

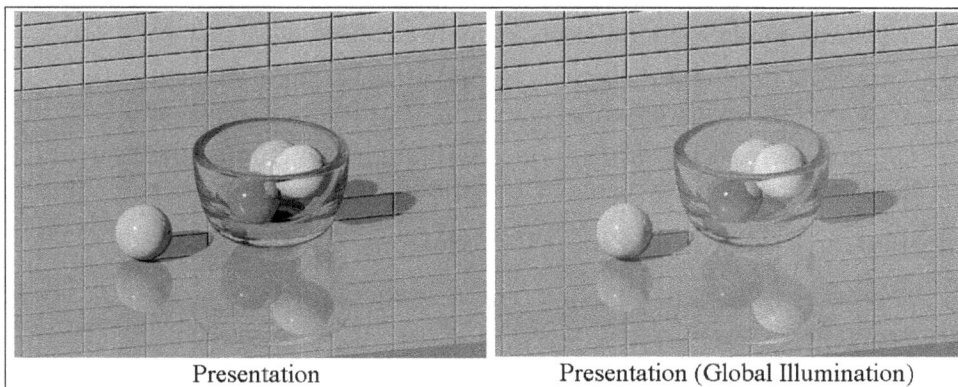

| Presentation | Presentation (Global Illumination) |

Scene illumination

We are going to see how to add light to scenes, making a distinction between natural lighting, from the Sun and sky simulation, and artificial lighting, with point lights, spot lights, distant lights, and web lights (with IES file information).

Default lighting

Before adding any light, AutoCAD applies default lighting, consisting of two distant lights parallel to the viewing direction. When orbiting the view, this default lighting follows. That's why if we make a render without applied lights, the model can be seen without shadows, which is unrealistic.

When we create the first light or activate the Sun light, AutoCAD prompts for turning off the default lighting. With lights applied, the DEFAULTLIGHTING variable controls which illumination will be used in the current viewport: 0 to applied lights and 1 to default lighting. This variable does not affect renderings.

Generic and photometric lighting

AutoCAD can calculate indirect lighting, thus adding realism to the scene, but it takes more time to render. We achieve this by applying photometric lights, whose intensities are measured in real lighting units, such as candela or foot-candles.

Three options are available:

- **International**: Photometric lights are applied and light intensities are measured in international units, namely candela
- **American**: Photometric lights are applied where light intensities are measured in American units, namely foot-candles
- **Generic**: Generic lights are applied without lighting units, and each light has a generic intensity value of 1, without correspondence to real lights

Illumination can be activated by using the following:

- The UNITS command (alias UN): This is available on the **Lighting** area of the dialog box
- The **Lights** panel: This is available on the **Render** tab in the **Lights** panel
- The LIGHTINGUNITS variable: This variable, which is saved with the drawing, controls the **Generic** (value 0), **American** (value 1), and **International** (value 2) lighting units

The options can be seen in the following screenshot:

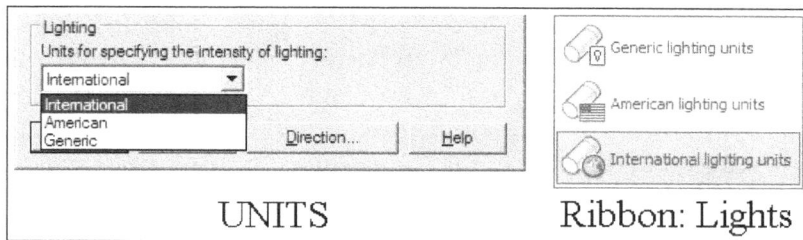

UNITS   Ribbon: Lights

> With generic lighting, renders are faster; however, as it does not calculate indirect lighting, to get some realism we need to place several "false" lights (without shadows) to simulate light reflecting on objects.

Sun light and sky light

To simulate light from the Sun, we apply the next command. With photometric lights (**International** or **American**), we can also turn on lighting simulating the sky (coming from all around the model). For defining the position of the Sun related to the model (azimuth and altitude), the geographic location, north direction, hour, and date must be defined.

The SUNPROPERTIES command

The SUNPROPERTIES command (no alias) displays a palette with all the parameters to define the Sun light. This light is infinite and projects parallel shadows. On the palette, displayed in the following image, we have the following areas:

- **General**: **Status** allows turning the Sun light on or off. With **Intensity**, we can increase or decrease the Sun light's intensity. **Color** controls the Sun light color in generic lighting; with photometric lighting, color is automatically calculated based on location, date, and time. **Shadows** can be turned on or off.

- **Sky Properties**: This area is only available in photometric lighting and will be presented next.

- **Sun Angle Calculator**: With **Date** we choose the day, and with **Time** the precise hour and minute of the day to calculate the Sun light. With the **Daylight Saving** option, we can adjust the solar hour considering economical reasons. The remaining values (**Azimuth**, **Altitude**, and **Source Vector**) are calculated from previous parameters and geographic locations.

- **Rendered Shadow Details**: **Type** controls the shadow type between **Sharp**, **Soft (mapped)**, and **Soft (area)**. With generic lighting, we can choose between sharp and soft shadow edges. With photometric lighting, **Type** is always **Soft (area)**, controlling **Samples** (precision) and **Softness** (detail of the edges).

- **Geographic Location**: This panel displays the current geographic location. The small icon to the right of the **Geographic Location** tab starts the GEOGRAPHICLOCATION command, which is presented in the next topic.

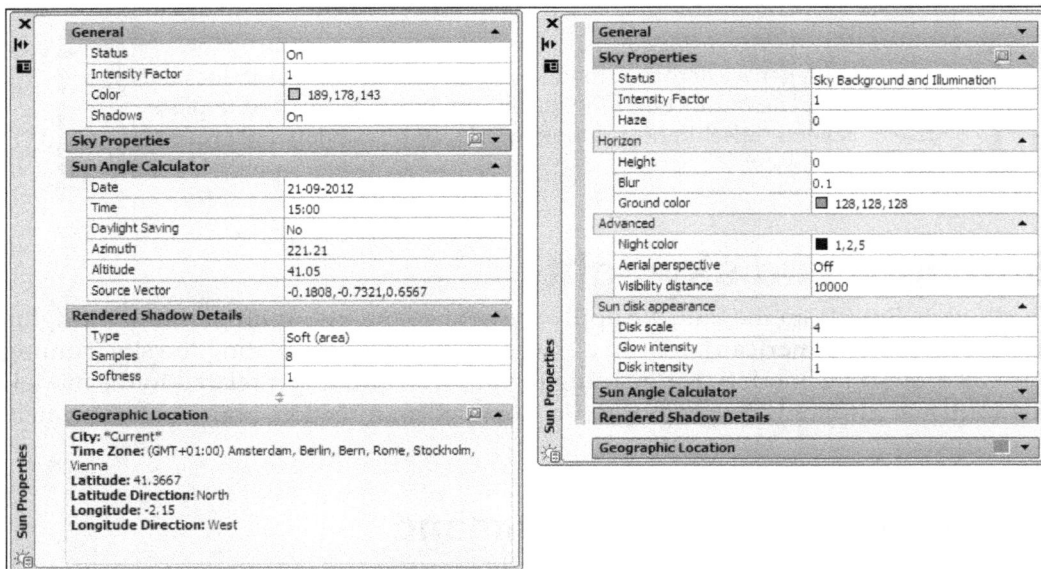

General		▲
Status	On	
Intensity Factor	1	
Color	☐ 189,178,143	
Shadows	On	

Sky Properties		🔲 ▼

Sun Angle Calculator		▲
Date	21-09-2012	
Time	15:00	
Daylight Saving	No	
Azimuth	221.21	
Altitude	41.05	
Source Vector	-0.1808,-0.7321,0.6567	

Rendered Shadow Details		▲
Type	Soft (area)	
Samples	8	
Softness	1	

Geographic Location 🔲 ▲
City: "Current"
Time Zone: (GMT+01:00) Amsterdam, Berlin, Bern, Rome, Stockholm, Vienna
Latitude: 41.3667
Latitude Direction: North
Longitude: -2.15
Longitude Direction: West

Sun Properties

General		▼

Sky Properties		🔲 ▲
Status	Sky Background and Illumination	
Intensity Factor	1	
Haze	0	

Horizon		▲
Height	0	
Blur	0.1	
Ground color	☐ 128,128,128	

Advanced		▲
Night color	■ 1,2,5	
Aerial perspective	Off	
Visibility distance	10000	

Sun disk appearance		▲
Disk scale	4	
Glow intensity	1	
Disk intensity	1	

Sun Angle Calculator		▼

Rendered Shadow Details		▼

Geographic Location		■ ▼

Sun Properties

Sky properties

This panel, available with photometric lighting, controls the sky background and sky illumination. The small icon to the right of the **Sky Properties** tab displays the **Adjust Sun & Sky Background** dialog box with similar parameters as this panel and with a preview area. **Status** allows us to activate sky lighting and/or the sky background, **Intensity Factor** allows us to increase or decrease the sky lighting intensity, and **Haze** defines atmospheric effects.

On **Horizon**, we can define the horizon line and ground plane, namely, its absolute **Height**, horizon **Blur**, and **Ground Color**. On **Advanced**, we define the sky color at night, aerial perspective, and visibility distance. Finally, on **Sun Disk Appearance**, we specify the Sun visualization, namely, **Disk scale**, **Glow intensity**, and **Disk intensity**.

The **Sun & Sky** background can only be used with the perspective view. With the perspective view, if **Status** is **Off** and cannot be modified, we may go to the **Adjust Sun & Sky Background** box and turn it on there.

| 13h00 | 18h00 |

The GEOGRAPHICLOCATION command

The GEOGRAPHICLOCATION command (alias GEO) defines a geographic location and the north direction, essential to calculate azimuth and altitude for the Sun. The first time we apply the command, a window is displayed prompting us to choose between three possibilities:

- **Import a .kml or .kmz file**: This imports the latitude, longitude, and elevation values from a KML or KMZ file. This file is an XML notation that includes geographic information created for Google Earth.

- **Import the current location from Google Earth**: If Google Earth is open, it imports the current location from Google Earth directly to the scene.

- **Enter the location values**: We choose the location and the north direction on a new dialog box, as follows:

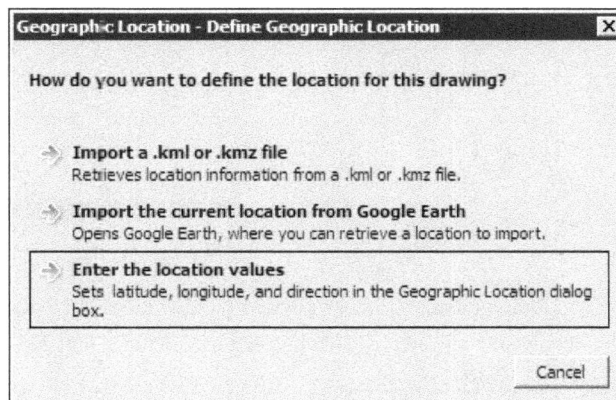

If we already have a defined geographic location, a similar box is displayed with options to edit the current location, to define a new one, or to delete it.

When creating or editing a geographic location, a new box is displayed with the following areas (displayed in the following image):

- **Latitude & Longitude**: The first list allows us to choose latitude and longitude coordinates between decimal values, or degrees / minutes / seconds. In **Latitude**, we enter latitude value, between -90 degrees and +90 degrees, and related to **North** or **South**. In **Longitude**, we enter the longitude value, between -180 degrees and +180 degrees, and related to **West** or **East**. The **Use Map** button displays a new box to specify the location. This can be done directly on the map displayed by selecting from **Region**, or from the **Nearest City** list. The **Time Zone** list allows us to choose the time zone corresponding to the chosen location. If **Nearest Big City** is on, when picking on the map, the location jumps to the nearest city.

- **Coordinates and elevation**: In this area, we specify the drawing coordinates and elevation where the geographic location mark will be placed.

- **North direction**: We define the north direction, related to the Y axis of the **world coordinate system (WCS)**. If not modified, the north direction is given by the Y axis of the WCS.

- **Up direction**: We define which axis of the world coordinate system is vertical.

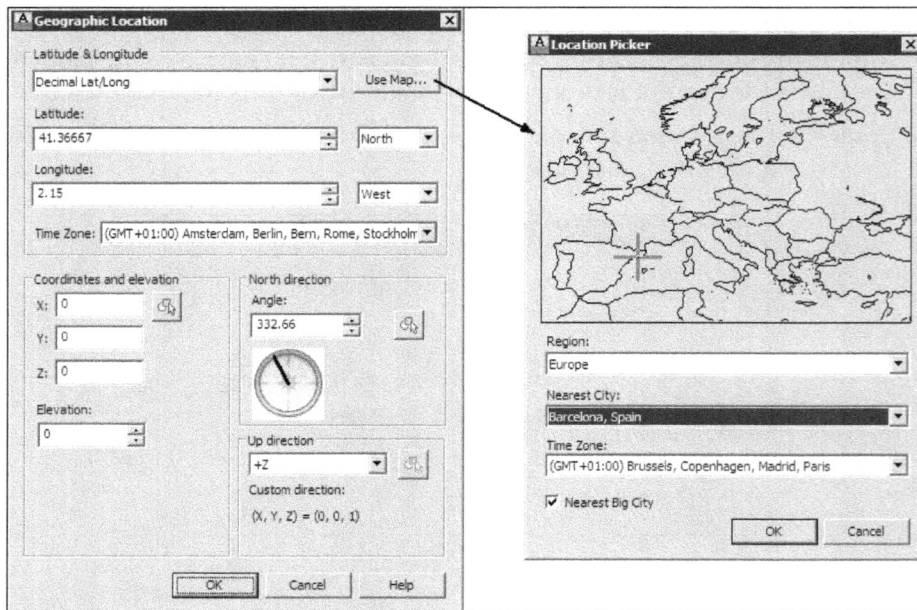

> To orient our scene related to the Sun, it is much better to change **North direction** than to rotate the entire scene. When changing **North direction**, **ViewCube** rotates accordingly and its top view is different from the plan view.
>
> The LATITUDE and LONGITUDE variables indicate the current values of latitude and longitude. The NORTHDIRECTION variable indicates the north direction related to the Y axis of the world coordinate system.

Exercise 10.2

We will render a 3D model at two different times of the day:

1. We open the drawing A3D_10_02.DWG of a simple house.

2. Applying a render, we will see the model, but only with the default lighting.

3. With SUNPROPERTIES, we turn the Sun light on in **Status**. We choose 09:00 in **Time** and render again. We maintain the date 21-09-2012. We get shadows but the result is not very good, as they are completely black.

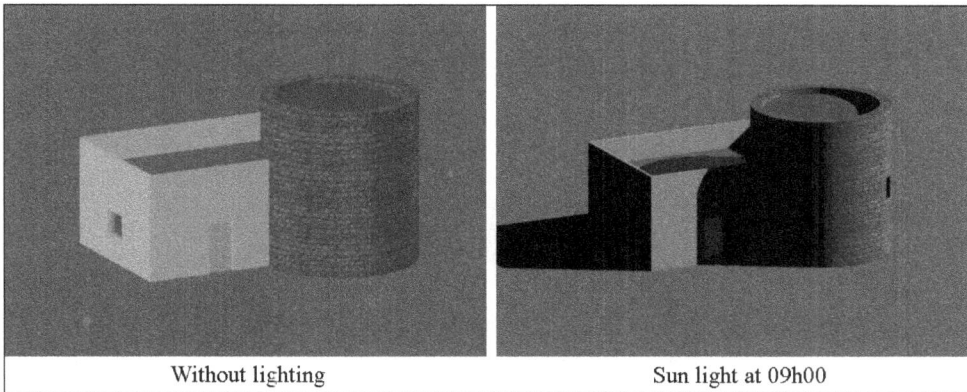

| Without lighting | Sun light at 09h00 |

4. Let's activate photometric lighting to get sky light. With the UNITS command, in **Lighting** we select **International**. The **Sun Properties** palette displays the **Sky Properties** panel.

5. However, the **Status** option in the **Sky Properties** panel is **Off** and cannot be modified. We need to set 1 as the PERSPECTIVE variable and zoom out to see the model. If the mouse wheel increment is too much, we apply the **Realtime** option after using the ZOOM command. When we click on the small icon on the right of the **Sky Properties** panel now, we finally get the **Sky Background and Illumination** option on the window that opens.

6. Now, we are going to choose the location. Still on the **Sun Properties** palette, we pick the **Geographic Location** button and confirm **Enter the location values**.

7. Through **Use Map**, we specify Barcelona's location and time zone (GMT +01:00) and confirm the time zone modification and the geographic location.

8. Finally, we get renders at **09h00** and **18h00**, and the quality is much better.

Sun light at 09h00   Sun light at 18h00

9. If we rotate the view to the right, we can see and render the solar disk.

10. We save the drawing with the name A3D_10_02final.DWG.

Point lights

We start artificial lighting with point lights. Besides the POINTLIGHT command, AutoCAD also provides a target point light. Point lights are represented by a small spherical glyph.

The POINTLIGHT command

The POINTLIGHT command (no alias) creates a point light in the drawing, that is, a light that has a specific location and radiates light in all directions. Examples are a bulb, a fire, or a candle.

The command prompts for the light location are as follows:

```
Command: POINTLIGHT
Specify source location <0,0,0>: Point
```

We press the *Enter* key to exit the command or choose an option:

```
Enter an option to change [Name/Intensity factor/Status/Photometry/
shadoW/Attenuation/filterColor/eXit] <eXit>: Enter or option
```

The command has the following options:

- **Name**: This allows us to give a name to the light, which is quite useful when having many lights.

- **Intensity factor**: This allows us to specify an intensity or brightness with a value starting at zero.

- **Status**: This allows us to turn the light on or off.

- **Photometry**: This option is only available with photometric lighting and allows us to specify the intensity and color in lighting units, such as candela, lux, or foot-candle. **Color** allows us to specify the color of the light based on a name or temperature in Kelvins.

- **shadoW**: This allows us to turn the shadows' projection on or off and select its type to be sharp or soft.

- **Attenuation**: With generic lighting, this option allows us to define attenuation with distance for the light. The **attenuation Type** option controls if there is no attenuation (**None**), if it's linear (**Inverse linear**), or if it's quadratic (**Inverse Squared**). With **Use limits**, we define two distances for starting and for ending. With photometric lighting, attenuation is set to **Inverse Squared** and cannot be modified.

- **filterColor**: This controls light color, allowing us to choose between a true color (red, green, blue values separated by commas), an AutoCAD color, or a color book.

> To move point lights, we can move its glyph. If we want to control its position or any other parameter, the best way is to apply the PROPERTIES command.
>
> The PROPERTIES command also allows us to change the type of light.

The TARGETPOINT command

The TARGETPOINT command (no alias) creates a special point light that emits light at a specific direction. This command can only be accessed from the command line. The command prompts for the light location and light target and displays the same options as the previous command:

```
Command: TARGETPOINT
Specify source location <0,0,0>: Point
Specify target location <0,0,-10>: Point
Enter an option to change [Name/Intensity factor/Status/Photometry/
shadoW/Attenuation/filterColor/eXit] <eXit>: Enter or Option
```

Spot lights

Other important artificial lights are spot lights, which simulate projectors, spots, or flashlights. As with point lights, we have two commands, the first one with a target and the second one without a target. Spot lights are represented by a small flashlight glyph.

The SPOTLIGHT command

The SPOTLIGHT command (no alias) allows us to create a light that projects in the shape of a cone. The command prompts for its position and target (where the light points to) are as follows:

```
Command: SPOTLIGHT
Specify source location <0,0,0>: Point
Specify target location <0,0,-10>: Point
```

We press *Enter* to exit the command or choose an option:

```
Enter an option to change [Name/Intensity factor/Status/Photometry/
Hotspot/Falloff/shadoW/Attenuation/filterColor/eXit] <eXit>: Enter or
option
```

The options for this command are as follows; most are the same as those for point lights:

- **Name**: This allows us to give a name to the light.

- **Intensity factor**: This allows us to specify an intensity or brightness, with a value starting from zero.

- **Status**: This allows us to turn the light on or off.

- **Photometry**: This option is only available with photometric lighting and allows us to specify the intensity and color in lighting units, such as candela, lux, or foot-candle. **Color** allows us to specify the color of the light based on a name or temperature in Kelvins.

- **Hotspot**: This specifies the cone angle of constant intensity, with a value ranging from 0 degrees to 160 degrees.

- **Falloff**: This specifies the total cone angle, including fading intensity, with a value ranging from 0 degrees to 160 degrees. It is never less than the **Hotspot** value. Outside this cone angle, there will be no light.

- **shadoW**: This allows us to turn the shadows' projection on or off and select its type to be sharp or soft.

- **Attenuation**: With generic lighting, it allows defining attenuation with distance for the light. The **attenuation Type** option controls if there is no attenuation (**None**), if it's linear (**Inverse linear**), or quadratic (**Inverse Squared**). With **Use limits**, we define two distances for starting and for ending. With photometric lighting, attenuation is set to **Inverse Squared** and cannot be modified.

- **filterColor**: This controls light color, allowing us to choose between a true color (red, green, and blue values, separated by commas), an AutoCAD color, or a color book.

> To adjust spot lights, including position, target or hotspot, and falloff angles, we can move its glyph grips. As with the previous lights, the PROPERTIES command includes all the parameters and also allows us to change the type of light.

The FREESPOT command

The FREESPOT command creates a special spot light, but without a target. To orient this light we may use the ROTATE3D command. FREESPOT can only be accessed from the command line. The command prompts for the light location and displays the same options as the previous command:

```
Command: FREESPOT
Specify source location <0,0,0>: Point
Enter an option to change [Name/Intensity factor/Status/Photometry/
Hotspot/Falloff/shadoW/Attenuation/filterColor/eXit] <eXit>: Enter or
option
```

Other lights

AutoCAD considers three more lights, one distant, which emits parallel rays, and two web lights, whose realistic intensities are derived from IES files.

The DISTANTLIGHT command

The DISTANTLIGHT command (no alias) creates a light that emits parallel rays, simulating a light that is far away. The most well-known example is the Sun. Since the SUNPROPERTIES command is available, this type of light is normally not used. These lights are not represented by glyphs.

The command prompts for a direction vector, which can be specified by two points:

```
Command: DISTANTLIGHT
Specify light direction FROM <0,0,0> or [Vector]: Point
Specify light direction TO <1,1,1>: Point
```

We press *Enter* to exit the command or choose an option:

```
Enter an option to change [Name/Intensity factor/Status/Photometry/
shadoW/filterColor/eXit] <eXit>: Enter or option
```

The **Vector** option specifies the lighting direction as a vector decomposed in the X, Y, and Z directions. For instance, a vector of 0,0,-1 means light is coming vertically from top to bottom. The remaining options are similar to the previous lighting commands.

> Application of these lights is not recommended with photometric lighting. But if you still do so, upon entering the command, a box requesting confirmation is displayed.
>
> As these lights don't have glyphs, the best way to select them is by the LIGHTLIST command, presented later in this chapter. Then, the PROPERTIES command includes all the parameters.

The WEBLIGHT command

The WEBLIGHT command (no alias) creates a web light. It is only available with photometric lighting. These web lights represent the light intensity distribution measured around the light by manufacturers, thus being much more accurate than other lights. The distribution information is provided in IES files. The command prompts for a light location and a target location are as follows:

```
Command: WEBLIGHT
Specify source location <0,0,0>: Point
Specify target location <0,0,-10>: Point
```

We press the *Enter* key to exit the command or choose an option:

```
Enter an option to change [Name/Intensity factor/Status/Photometry/
weB/shadoW/filterColor/eXit] <eXit>: Enter or option
```

The **weB** option displays several additional options: **File** specifies an IES file and **X**, **Y**, and **Z** allow us to rotate the light around the respective axis. The remaining options are similar to the previous commands.

> On the **TOOLPALETTES** palette, some web lights are included and the respective IES files come with AutoCAD.
>
> With a web light selected, the PROPERTIES command includes a **Photometric Web** panel specifying the IES file and displaying the intensity diagram, and a **Web offsets** panel specifying the rotations around the X, Y, or Z axis.

The FREEWEB command

The FREEWEB command (no alias) creates a web light, but without specifying a light target. All the remaining parameters are identical to the previous commands.

Other lighting commands

Also included are commands for selecting and managing lights, for controlling exposure (thus brightening and darkening the render image), and for converting lights from versions prior to 2007.

The LIGHTLIST command

The LIGHTLIST command (no alias) allows us to select lights by name. It displays the **Lights in Model** palette with the name of all lights in the scene, including distant lights. The only light not included is the Sun light. Selecting one or more lights on this list selects them in the scene, and on right-clicking we can delete the light, display their parameters on the **PROPERTIES** palette, or control the glyph display.

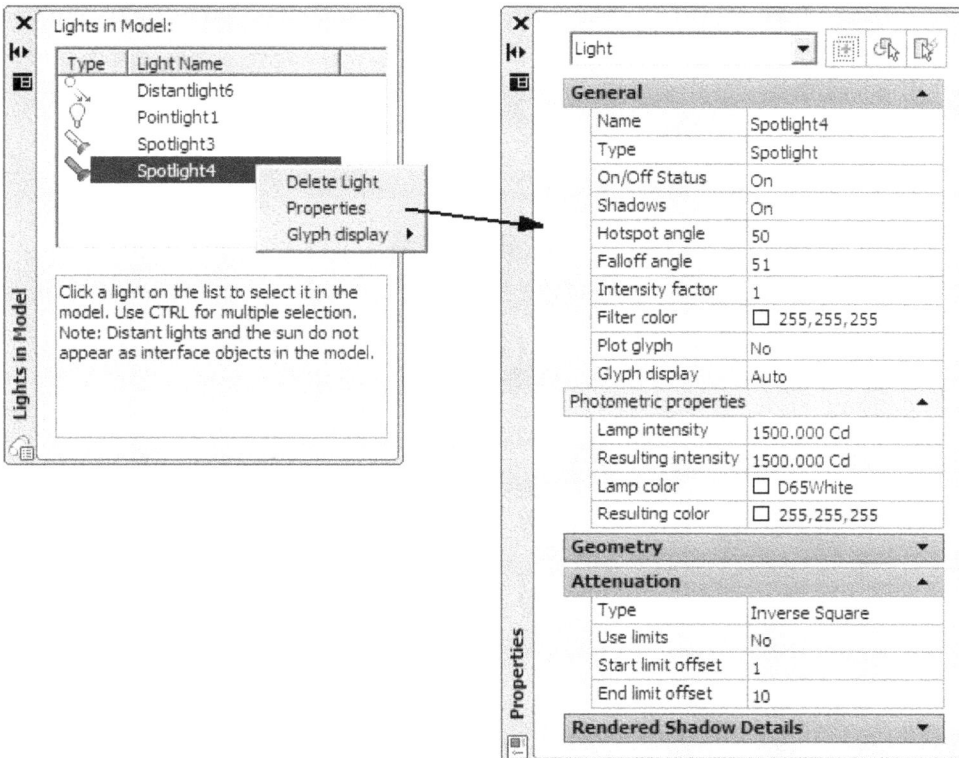

The RENDEREXPOSURE command

The RENDEREXPOSURE command (no alias) controls the exposure when rendering. This command is not available in generic lighting. It displays the **Adjust Rendered Exposure** dialog box with a **Preview** area that updates immediately and has the following options:

- **Brightness**: This adjusts the brightness value applied to the render, with a value ranging from 0 to 200. It is 65 by default.

- **Contrast**: This adjusts contrast, with a value ranging from 0 to 100. It is 50 by default.

- **Mid tones**: This adjusts mid-tones, that is, brightening or darkening intermediate colors without modifying most bright or dark colors. It accepts a value from 0 to 20; default is 1.

- **Exterior Daylight**: This controls whether exterior daylight is considered in the exposure or not.

- **Process Background**: This controls whether the applied background is considered in the exposure or not.

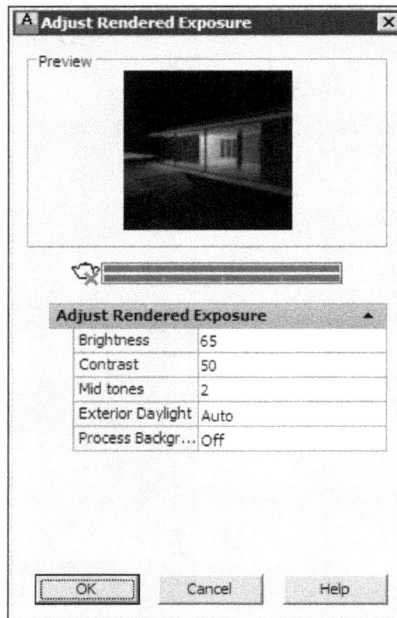

The following image includes an application of this command:

| Brightness 65 | Brightness 70 |
| Mid tones 2 | Mid tones 4 |

The CONVERTOLDLIGHTS command

If we open a 3D drawing from versions prior to 2007, the existing lights are not compatible with current lighting. The CONVERTOLDLIGHTS command (no alias) allows us to convert all these lights to current lights. The command is automatic and does not prompt for nothing.

Exercise 10.3

From the 3D model used in the *Exercise 10.2* section, we will now apply artificial lighting inside this house:

1. We open the drawing A3D_10_03.DWG.

2. With SUNPROPERTIES, we take 20h00 as the time. A render displays a completely black night.

3. To place lights inside, it is useful to change the visual style to 2D wireframe. Inside the house, we can see some small cylinders on the top. That is where we are going to place lights.

4. We apply the first point light on the center of the base of the left cylinder that is far from the door. With the VIEW command, we activate the view called **Camera1**.

5. Doing a render still displays black. So, we increase the light's **Intensity factor** to 100. A render already shows some light, but there is no indirect lighting, that is, light does not reflect on walls and other surfaces.

6. With the RPREF command, we turn on **Global Illumination** by clicking on the small lamp icon. A render looks completely different now:

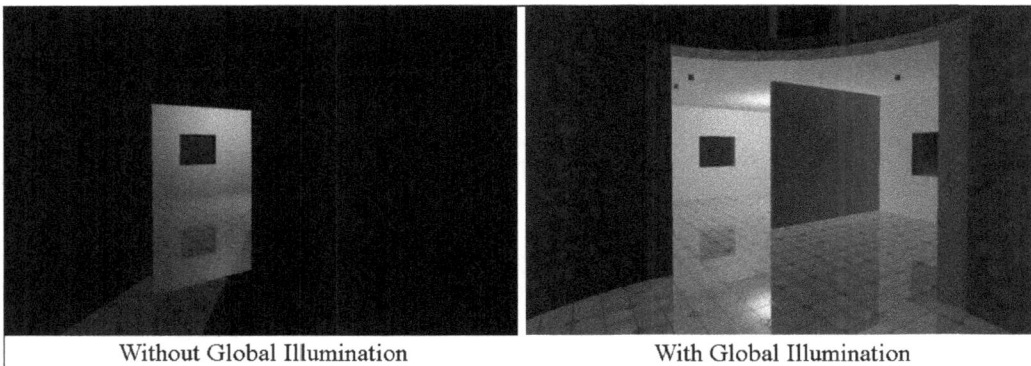

| Without Global Illumination | With Global Illumination |

7. We copy the light to the right cylinder. Now a render shows too much light. With the `LIGHTLIST` command, we select both point lights and reduce **Intensity factor** to `50`:

8. On the base of the cylinders, near the door, we apply spot lights. We create the first one at the center of the base, and when prompted for the target we write `@0,0,-2.5` as the absolute coordinates of the center and give **Hotspot** angle as `70` and **Falloff** angle as `100`.

9. We copy this light to the base of the other two cylinders. Again, with `LIGHTLIST`, we select the three spot lights and increase **Intensity value** to `50`. If we turn off the point lights, we will see only the spotlights' illumination.

10. We turn on the point lights again and render a final scene.

| Only spot lights | Spot and point lights |

11. We save the drawing with the name `A3D_10_03final.DWG`.

Summary

This chapter includes all commands and concepts related to rendering and lighting. Rendering allows us to obtain photorealistic images from our 3D scene.

General concepts and a typical workflow were presented, followed by the RENDER command, the main command to render, and also save the render image. Other related commands are: RENDERWIN to display the render windows; RENDERCROP to render just some part of the viewport, and is useful for testing; SAVEIMG to save a rendered image in viewport; RPREF to specify all default parameters used for rendering, including activation of global illumination and final gather; and RENDERPRESETS to create rendering presets.

We then introduced illumination, starting with default, generic, and photometric lighting. To apply the natural lighting of the Sun, there is the SUNPROPERTIES command, which can also consider sky lighting and background. GEOGRAPHICLOCATION specifies the location and the north direction. There are several commands to apply artificial lighting: POINTLIGHT to create point lights, SPOTLIGHT to create spot lights, DISTANTLIGHT to simulate lights that are far away and project parallel rays, and WEBLIGHT to consider web lights coming from IES files. To manage lights, we apply the LIGHTLIST command. When using photometric lighting, the RENDEREXPOSURE command allows us to control the exposure. Finally, when opening scenes from versions prior to 2007, the CONVERTOLDLIGHTS command converts these old lights.

In the next chapter we will explore materials and effects.

11
Materials and Effects

After seeing render and illumination in the last chapter, in this chapter we will complete the render subject with materials and effects. As important as lighting a 3D scene is the application of realistic materials that resemble materials of the real world. AutoCAD also allows us to specify scene backgrounds and applying fog effect.

The topics covered in this chapter are:

- Understanding materials
- How to apply materials
- How to open a material library
- How to create materials from different templates
- How to create textures
- How to control mapping coordinates
- Applying backgrounds and fog

Introduction to materials, textures, and effects

Since Version 2011, with the introduction of Autodesk Materials and the inclusion of more than 700 excellent looking and ready-to-apply materials, photo-realistic images can easily be obtained. These materials can be adjusted and we can also create countless materials, generic or template-based.

Materials are defined from several parameters, often including textures or maps. Textures can be photos, images, and also several procedural maps simulating tiles, chess patterns, wood, marble, and others. When using textured materials, mapping coordinates (the way textures are painted on faces, including texture origin, scale, and rotation) can be adjusted.

Until applying materials, a special material, called Global with color defined by the object, is applied to all objects. This material cannot be deleted or renamed. Materials can be applied to objects, to faces of objects or by layer, the last being the preferable way.

As effects, AutoCAD only includes background and fog or depth cue.

Materials and textures

Corrected materials and respective textures are essential for good renders. AutoCAD provides several commands for selecting, creating, and applying materials.

Autodesk Material Library

Starting with Version 2011, when installing AutoCAD, the Autodesk Material Library is also installed. This material library is shared with other Autodesk software, as 3ds Max or Revit, and can not be modified, but allows their materials to be loaded into the current drawing, which are changed and then saved in a user library.

The Autodesk Material Library containing more than 1,000 textures is defined in a folder similar to `C:\Program Files (x86)\Common Files\Autodesk Shared\ Materials\` (depending on the Windows version and language).

Under this folder, textures (images or photos) are available in three groups:

- `\assetlibrary_base.fbm\Textures\1\Mats`: More than 1,000 images with 256 x 256 resolution

- `\assetlibrary_base.fbm\Textures\2\Mats`: More than 1,000 images (Versions 2011/2012) and almost 100 images (Version 2013) with 512 x 512 resolution

- `\assetlibrary_base.fbm\Textures\3\Mats`: Almost 1,000 images with 1024 x 1024 resolution

These textures have an excellent quality and can also be used to create new materials.

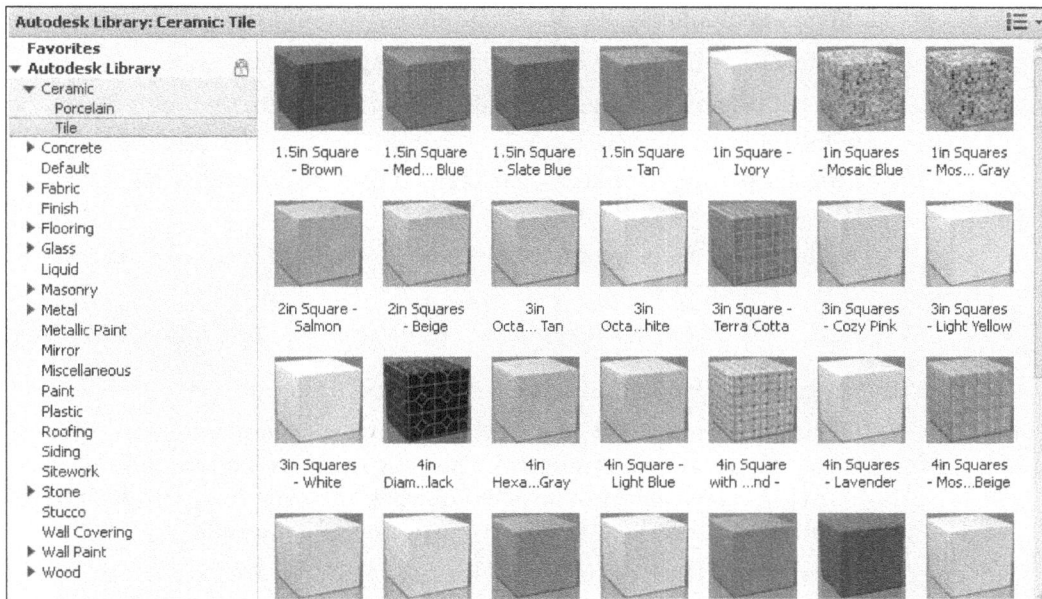

Managing materials and libraries

The main command for materials is MATBROWSEROPEN, in which we can load materials, assign to selection, and also access the material editing command.

The MATBROWSEROPEN command

The MATBROWSEROPEN command (alias MAT) displays the palette **Materials Browser**. This command can be found on the **Render** toolbar, on the ribbon, the **Render** tab, and on the **View/Render** menu bar. The palette includes the following areas and buttons, as displayed in the next screenshot:

- **Search**: This area allows searching a material in all open libraries. Upon writing, the available materials are filtered.

- **Document Materials: All**: This area displays the loaded materials in the current drawing. By selecting a material and right-clicking (shortcut menu), some options are available:

    ○ **Assign to Selection**: This assigns the material to the selected objects

    ○ **Select Objects Applied To**: This selects all objects that own this material

    ○ **Edit, Duplicate, Rename**, and **Delete**: These allow to, respectively, edit, make a copy, rename, or delete the material

    ○ **Add to**: This adds the material to the Favorites library or to the current tool palette

    ○ **Purge All Unused**: This removes unused materials from the drawing

- **Document bar**: This list includes some filters and options:

    ○ **Document Materials**: This contains filters for displaying loaded materials, namely showing all, showing only applied materials, showing materials from selected objects, showing loaded but unused materials, and an option to purge all unused materials

    ○ **View Type**: This controls how loaded materials are displayed

    ○ **Sort**: This controls the order for displaying materials

    ○ **Thumbnail Size**: This controls the size in pixels for material thumbnails

- **Library**: This area includes, on the left-hand side, all libraries and respective categories. Besides user libraries, the Autodesk Library and Favorites are always available. On the right-hand side, all materials of the selected category or library are displayed.

- **Library bar**: This list includes some filters and options:
    - ° **Library**: This displays the selected library
    - ° **Hide Library Tree**: This hides the left-hand area, thus maximizing the materials view
    - ° **View Type**: This controls how materials are displayed
    - ° **Sort**: This controls the order for displaying materials
    - ° **Thumbnail Size**: This controls the size in pixels for material thumbnails

- **Manage**: This button displays options related to libraries:
    - ° **Open Existing Library**: This opens a user library, displaying a file dialog box to specify the ADSKLIB file
    - ° **Create New Library**: This creates a user library, displaying a file dialog box to specify a location and name for the ADSKLIB file
    - ° **Remove Library**: This removes a user library
    - ° **Create Category**: This creates a category for a user or favorite libraries
    - ° **Delete Category**: This erases categories for the same libraries
    - ° **Rename**: This renames user libraries or categories

- **Create Material**: This button displays a list of available templates to create a material, such as **Ceramic**, **Concrete**, **Wall Paint**, or **Wood**. The **Generic** template allows creating any material. The material creation is presented later.

- **Materials Editor**: This button accesses **Material Editor** (using the `MATEDITOROPEN` command) for the selected material or the global material if none is selected. This command is presented later.

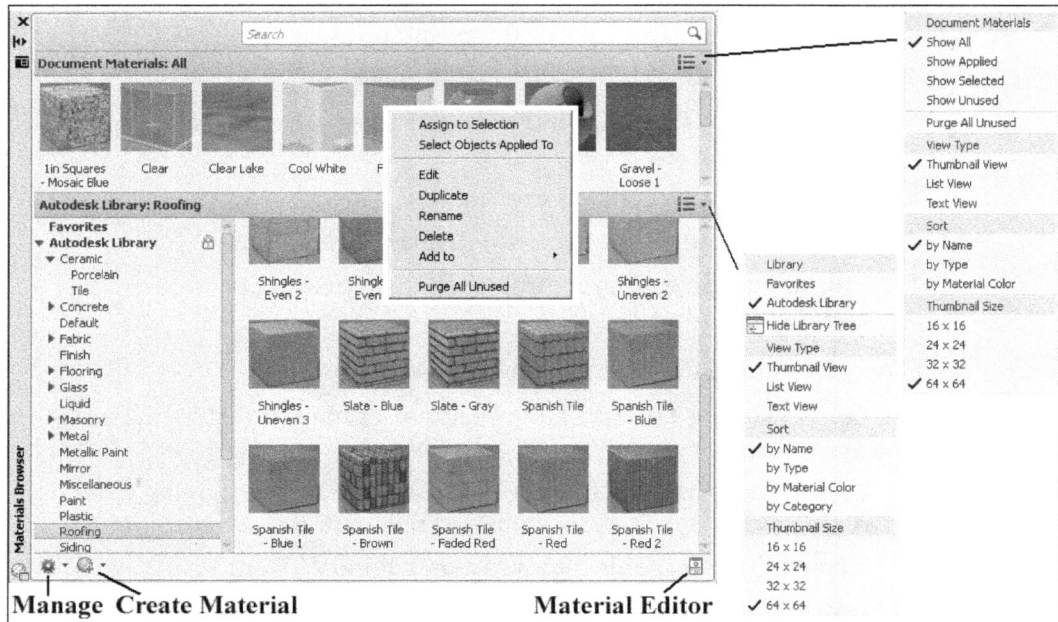

To load a material to the current drawing, we can double-click on the name or preview of a material, or drag it from the library to the **Document Materials** area.

> Before loading materials, it is wise to check the `UNITS` command (alias `UN`). The **Insertion scale** unit should be according to our model, so texture sizes are nearer to what we need and immediately visible when rendering or with the **Realistic visual** style.
>
> When passing over a material from a library, two small icons are displayed. The left one loads the material and the right one, besides loading it, also opens it on **Materials Editor**. But be careful; both icons apply the material directly to selected objects.

Assigning materials

The MATERIALATTACH command allows you to assign commands by layer, thus being the most flexible way for assigning materials, even to objects not yet created. Also we may apply the MATERIALASSIGN command. Besides these two commands, there are some more ways to assign materials to objects or object faces:

- Dragging the material to the object: We can drag the material from the **Document Materials** area or the **Library** area to objects

- Selecting objects and applying the **Assign to Selection** option: We first select objects and then apply this option by right-clicking on the menu

- Selecting objects and applying the PROPERTIES palette: We select the objects and choose one of the loaded materials with the PROPERTIES palette, available at **3D Visualization/Material**

- Selecting subobjects and applying the **Assign to Selection** option: We first select subobjects with the *Ctrl* key and then apply this option by right-clicking on the menu

> Materials applied directly, without being by layer, take preference over the layer material, as it happens with color or linetype. To have control back by layer, we select these objects and choose **ByLayer** on the properties palette.
>
> When applying a new material to an object, the old one is removed. To remove a material without applying a new one, it is enough to apply the **Global** material.

The MATERIALATTACH command

The MATERIALATTACH command (no alias) allows you to apply materials by layer. It displays a dialog box, where we drag materials from the left-hand area to layer names on the right-hand area. Big red crosses allow removing materials from layers.

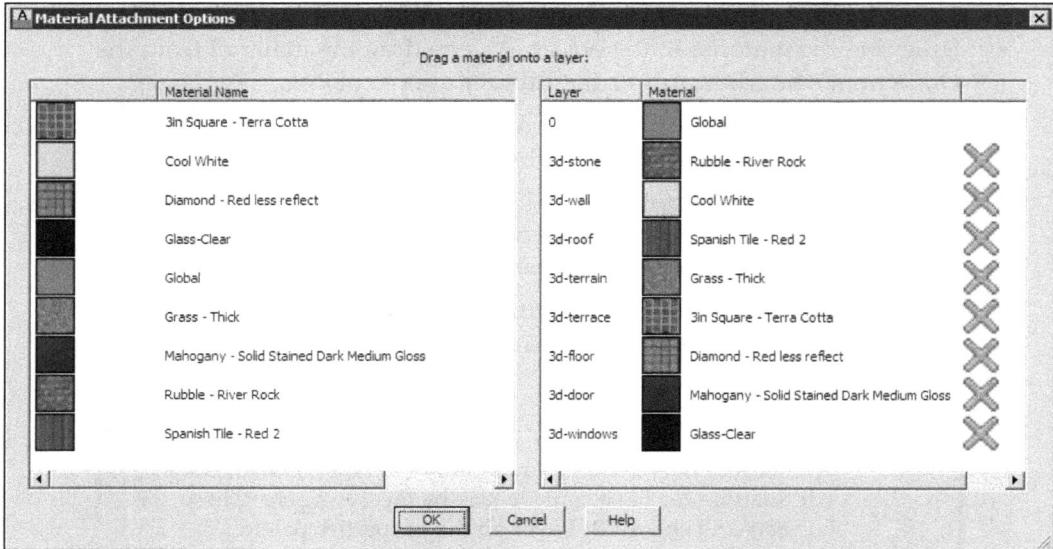

The MATERIALASSIGN command

The MATERIALASSIGN command (no alias) allows assigning the material defined in the CMATERIAL variable to selected objects. The command just prompts for object selection.

If specifying a material in the PROPERTIES palette without any selected object, we are defining a default material for new objects, whose interest is minimal.

Exercise 11.1

To the mechanical assembly drawing used in the documentation chapter, we are going to pick some materials and assign them by layer:

1. Open the drawing A3D_11_01.DWG, which includes a bearing. The lighting is already set up to photometric, with sunlight and skylight.

2. First, verify if the model unit is correct. Applying the UNITS command, we confirm that the unit is in millimeter.

3. With the MATBROWSEROPEN command (alias MAT), pick the required materials. In the **Metal** category of Autodesk Library, double-click on materials, such as **Steel**, **Steel-Cast**, **Machined 02**, and **Semi-Polished**. To the table, select, under the **Stone** category, **Green Polished**.

4. Applying the MATERIALATTACH command, assign **Steel-Cast** to **bottompart** layer, **Steel** to **toppart**, **Semi-Polished** to **screw**, **Machined 02** to **nut**, and **Green Polished** to **table**. Each material is dragged over the layer name.

5. Rendering the drawing (the RENDER command), see all applied materials, including the one applied to the table.

6. Save the drawing with the name A3D_11_01final.DWG.

Creating and editing materials

With the MATBROWSEROPEN command we can create materials from available templates. After specifying a template material, it immediately opens the **Materials Editor** palette, that is, it runs the MATEDITOROPEN command.

The MATEDITOROPEN command

The MATEDITOROPEN command (no alias) allows you to create and edit materials. It displays a palette with two tabs for the material selected in the loaded materials or for the new material. The **Information** tab displays information about the material name, type, and texture paths. We may modify the name, description, and keywords.

On the **Appearance** tab, besides parameters depending on the material template and presented next, there are some common areas and buttons:

- **Material preview and list**: This area can be resized, thus increasing or decreasing the preview size. The list allows for selecting the preview object, among 12 options, and its rendering quality, among four options.

- **Name**: We should give a specific and useful name to the material.

- **Create material**: This button displays a list of available templates to create a material, similar to the button used in the previous command.

- **Open/close Material Browser**: This button just allows to open or close the **Material Browser** palette.

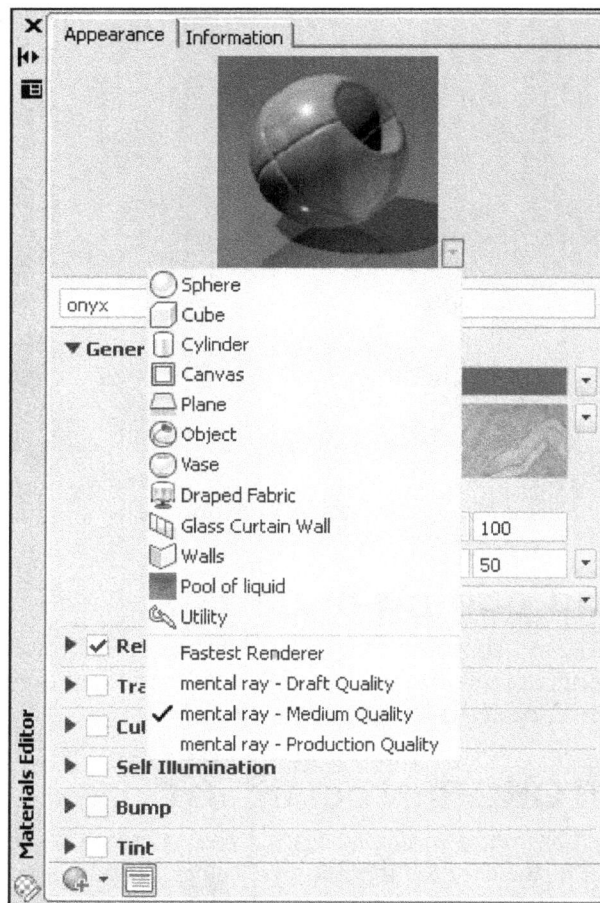

The Generic material

The **Generic** material template is the most complete template, allowing for the creation of any type of material.

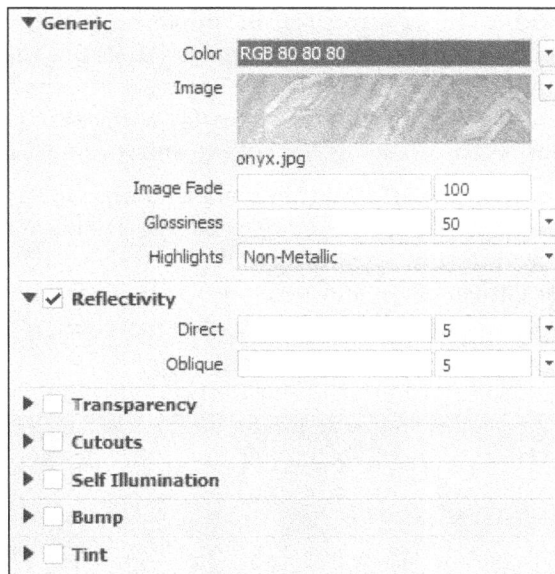

This material template has the following parameters:

- **Generic**: This parameter specifies the material base color, **Color** or diffuse texture, **Image**. When choosing or editing a texture, a new palette is displayed, which is presented next. The texture can be a photo, an image, or a procedural texture, calculated by AutoCAD at render time. **Image Fade** mixes the texture and base color. **Glossiness** controls the quality of reflective light; the smaller the highlight, the glossier the material. **Highlights** list specifies the type of specular highlights, between nonmetallic (highlight reflects the light color) and metallic (highlight maintains material color).

- **Reflectivity**: This controls the direct and oblique level of reflections.

- **Transparency**: This controls if and how transparent the material is. **Amount** defines the level of transparency (**0** is completely opaque and **100** is completely transparent). **Translucency** specifies a translucent material, that is, some light passes through and some light is scattered, thus reducing visibility. **Refraction** specifies the material IOR (index of refraction, a property for nonopaque materials). The list includes the most common IORs, such as glass (1.52) and water (1.33). To control opacity maps, it is better to apply cutout parameters instead of **Image** and **Image Fade**.

- **Cutouts**: We can specify a map to create an opacity map or cutout, that is, a material with different areas of transparency by selecting a map (image or texture). Black areas of the map are completely transparent, while white areas are completely opaque.

- **Self Illumination**: This controls self-illumination, simulating a lamp turned on or neon. With this parameter, the material does not cast light and neither receives any shadows. **Filter Color** simulates a filter applied to lighting color, **Luminance** simulates photometric intensity, with some predefined values, and **Color Temperature** adjusts the self-illumination color.

- **Bump**: This controls the effect of nonsmooth material, simulating irregularities such as the space between tiles or bricks. We can apply a map (image or texture) to simulate bump and adjust its value with the **Amount** option. Higher values mean a more irregular material. Lighter colors simulate high relief, darker colors simulate low relief.

- **Tint**: This specifies a color to tint the material.

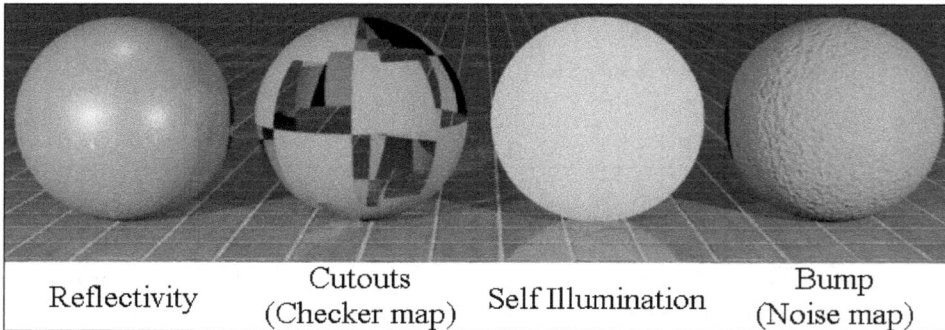

| Reflectivity | Cutouts (Checker map) | Self Illumination | Bump (Noise map) |

> A small amount of self-illumination can be applied to bright a material that looks too dark when rendering.

The Ceramic material

The **Ceramic** material template allows you to create materials simulating ceramic, porcelain, or tiles. **Type** specifies if it is ceramic or porcelain. **Image** allows selecting the image or procedural map. **Finish** indicates how glossy the material between **High Gloss/Glazed**, **Satin**, or **Mate** is. **Finish Bumps** controls the small bump waves specific to ceramics. **Relief Pattern** controls relief or main bump; in **Image** we choose an image similar to the main diffuse image but preferably gray-scaled, lighter areas will seem raised, and darker areas will seem lowered. The **Amount** value controls the relief importance. **Tint** specifies a color to tint the material as shown in the following screenshot:

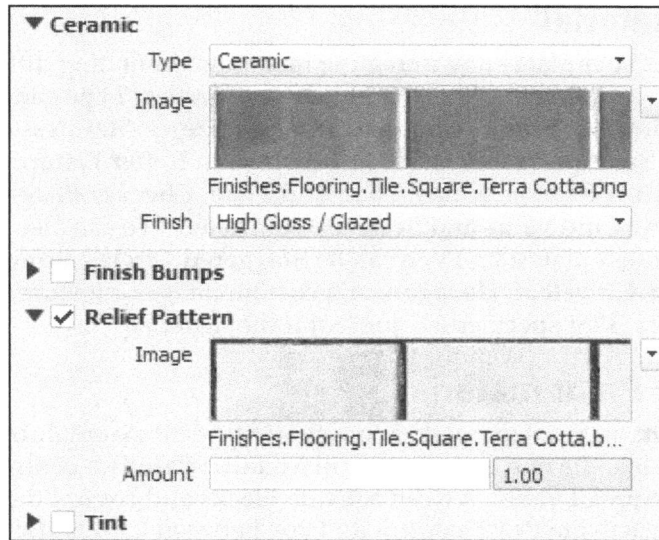

The Concrete material

The **Concrete** material template allows you to create materials simulating concrete. In **Concrete** we define the main image and whether **Sealant** is **Epoxy**, **Acrylic**, or **None**. In **Finish Bumps** we control the type of bump if it is **Broom Straight**, **Broom Curved**, **Smooth**, **Polished**, or **Stamped/Custom**; the last allowing type applying an image and control to its bump amount. **Weathering** allows applying discoloration automatically or based on an image. **Tint** specifies a color to tint the material.

Glazing material

The **Glazing** material template allows you to create materials simulating thin glass, where refraction is not considered. We control **Color**, value of **Reflectance**, and the number of **Sheets of Glass**, to a maximum of six.

The Masonry material

The **Masonry** material template allows you to create materials mainly to walls, simulating masonry, bricks, or **Concrete Masonry Unit** (**CMU**). In **Masonry**, the **Type** list defines masonry or CMU; **Image** allows choosing an image, **Finish** controls if the material looks **Unfinished**, **Glossy**, or **Matte**, and **Relief Pattern** controls relief or main bump. In **Image** we choose an image similar to the main diffused image but preferably gray-scaled; lighter areas will seem raised and darker areas will seem lowered. The **Amount** value controls the relief importance. **Tint** specifies a color to tint the material.

The Metal material

The **Metal** material template allows creating materials simulating different types of metals. In **Metal** we define the **Type** and **Finish** properties. **Type** can be **Aluminum**, **Anodized Aluminum**, **Chrome**, **Copper**, **Brass**, **Bronze**, or **Stainless Steel**. **Finish** can be **Polished**, **Semi-polished**, **Satin**, or **Brushed**. In **Relief Pattern** we control a bump effect; its **Type** can be **Knurl**, **Diamond Plate**, **Checker Plate**, or **Custom Image**. Also the **Amount** value and **Scale** are controlled. We can also apply **Cutouts**, simulating perforated plates; its **Type** can be **Staggered Circles**, **Straight Circles**, **Squares**, **Grecian**, **Cloveleaf**, **Hexagon**, or a custom image or texture; each type with related parameters. **Tint** specifies a color to tint the material.

The Metallic Paint material

The **Metallic Paint** material template allows you to create material simulating special paint over metal, like the one used in cars. In **Metallic Paint** we control the **Color** and a **Highlight Spread** value. We can activate **Flecks** and control their **Color** and **Size**. We can also activate **Pearl**, control its **Type** between **Chromatic** or **Second Color** and control its **Amount** value. In the paint **Top Coat**, **Type** can be **Car Paint**, **Chrome**, **Matte** or **Custom**, and **Finish** can be **Smooth** or **Orange Peel**. **Tint** specifies a color to tint the material.

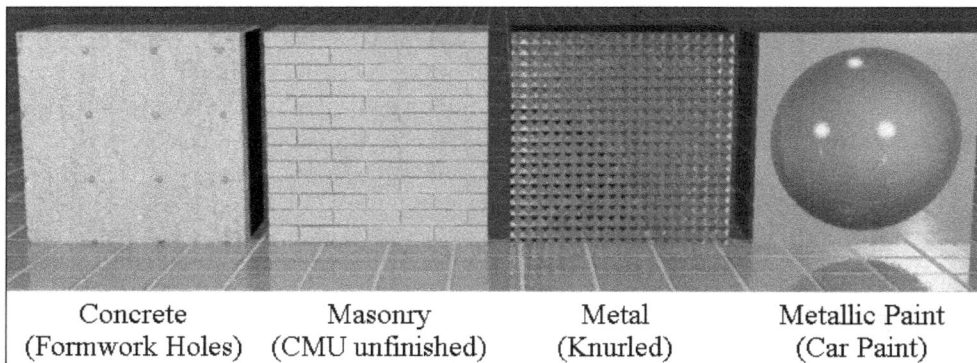

| Concrete | Masonry | Metal | Metallic Paint |
| (Formwork Holes) | (CMU unfinished) | (Knurled) | (Car Paint) |

The Mirror material

The **Mirror** material template allows creating materials simulating mirrors with full reflection. There is a unique parameter, **Tint Color**, to control the mirror color. **Tint** specifies a color to tint the material.

The Plastic material

The **Plastic** material template allows you to create materials simulating several types of plastic. In **Plastic**, **Type** can be **Plastic (Solid)**, **Transparent** or **Vinyl**, with a specified **Color**; **Finish** can be **Polished**, **Glossy** or **Matte**. **Finish Bumps** can be defined by an **Image** or texture with a specified **Amount**. **Relief Pattern** controls relief or main bump: in **Image** we choose an image; the **Amount** value controls the relief importance. **Tint** specifies a color to tint the material.

The Solid Glass material

The **Solid Glass** material template allows you to create materials simulating thick transparent materials, that is, considering refraction. In **Solid Glass**, we control **Color**, **Reflectance**, **Refraction**, and **Roughness**. With **Refraction**, we may select a material type from the list or enter the index of refraction value. **Relief Pattern** controls relief or main bump; its **Type** can be **Rippled**, **Wavy**, or **Custom**. The **Amount** value controls the relief importance. **Tint** specifies a color to tint the material.

The Stone material

The **Stone** material template allows you to create materials simulating stone. In **Stone**, **Image** allows choosing an image. **Finish** controls if the material looks **Polished**, **Unfinished**, **Glossy**, or **Matte**. **Finish Bumps** can be **Polished Granite**, **Stone Wall**, **Glossy Marble**, or **Custom** with a specified **Amount**. **Relief Pattern** controls relief or main bump. In **Image** we choose an image. the **Amount** value controls the relief importance. **Tint** specifies a color to tint the material.

| Mirror | Plastic | Solid Glass | Stone |

The Wall Paint material

The **Wall Paint** material template allows you to create materials simulating paint on a wall. In **Wall Paint**, we control **Color**. **Finish** can be **Flat/Matte**, **Eggshell**, **Platinum**, **Pearl**, **Semi-gloss**, or **Gloss**. The ink **Application** can be by **Roller**, **Brush**, or **Spray**. **Tint** specifies a color to tint the material.

The Water material

The **Water** material template allows you to create materials simulating water with only few parameters; its **Type** can be **Swimming Pool**, **Generic Reflecting Pool**, **Generic Stream/River**, **Generic Pond/Lake**, or **Generic Sea/Ocean**. In **Color** we have several predefined colors, including **Custom**. **Wave Height** controls the relative size of waves. **Tint** specifies a color to tint the material.

The Wood material

Finally, the **Wood** material template allows you to create materials simulating wood. In **Wood**, **Image** allows you to choose an image. **Stain** can be on or off. **Finish** controls if the material looks **Unfinished**, **Glossy Varnish**, **Semi-gloss Varnish**, or **Satin Varnish**. **Used For** specifies the material application if **Flooring** or **Furniture** is selected. **Relief Pattern** controls relief or main bump if **Based on Wood Grain** or **Custom** is selected. The **Amount** value controls the relief importance. **Tint** specifies a color to tint the material.

> The key to adjusting and adapting materials is trying. We may start from a material already included and proceed to some changes. Whenever we change a material, it is wise to give it a proper name, so as not to confuse it with the original.
>
> Instead of trying materials with complex models, it is better to fine-tune them with a sample object of similar size.

Images and other textures/maps

Most materials and material templates use images and textures, also known as maps, to compose the main aspect, cutouts, relief, and bump, and other parameters. Besides applying images, we can apply procedural maps to simulate tiles, wood, irregularities, and other effects. Procedural maps have the advantage of tiling perfectly adjusted effects, that is, effects are placed several times side by side, which look continuous, without a mosaic effect.

Every material parameter that accepts an image or texture/map displays a list containing all texture types, which are displayed in the next screenshot. After picking one, AutoCAD displays the **Texture Editor** palette with related parameters.

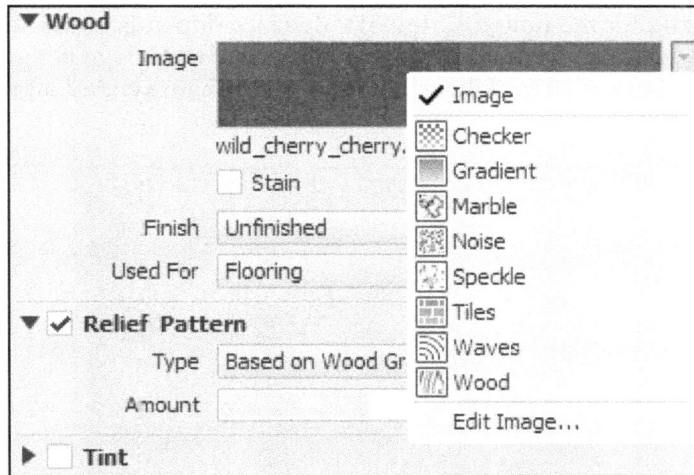

The procedural textures are similar to the same textures of 3ds Max, Revit, Showcase, and other Autodesk software.

Image

The **Image** map is the most used one that allows specifying a photo or image that represents the wanted texture to the material. AutoCAD accepts the image formats, such as JPG, PNG, TIF, BMP, EXR, HDR, DIB, PCX, GIF, TGA, and RLE. After selecting an image, the **Texture Editor** palette is displayed, with two panels. This palette is also displayed when editing an image:

- **Image**: In **Source** we select the image; **Brightness** controls how bright the image is and **Invert** allows inverting image colors.

- **Transforms**: With **Link texture transforms** checked, all transformations made in this panel are also applied to other images used by the material. **Offset** in the **Position** option controls the image origin related to faces and **Rotation** controls rotation. **Scale** controls real-world dimensions, width and height, for the image. Original value size depends on the current unit selected at UNITS command. **Repeat** controls if the image is repeated to fulfill the faces, that is, **Tile** or not. The small chain symbol allows locking both values.

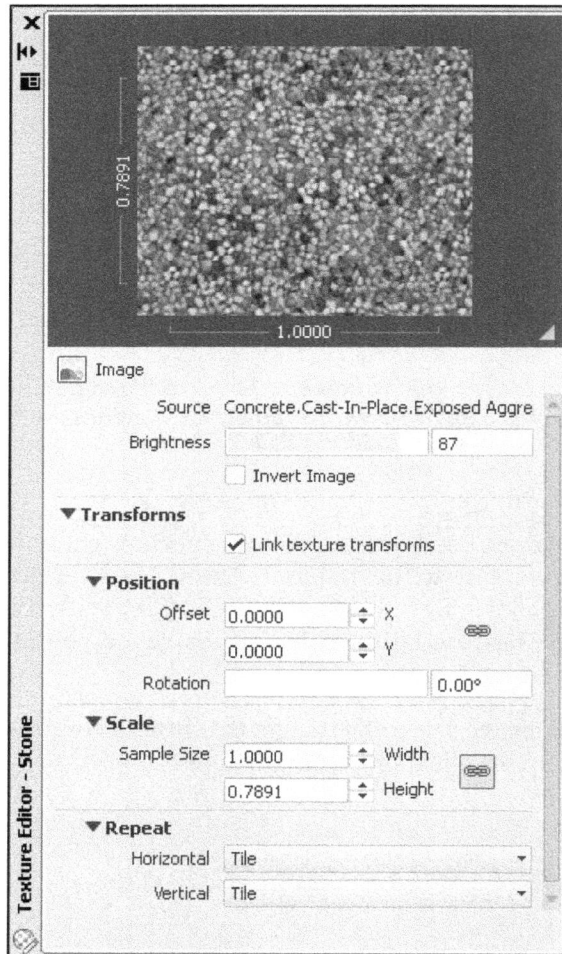

Image files can be dragged from Windows Explorer to texture slots on the **Materials Editor** palette.

Checker

The **Checker** texture simulates a pattern composed by alternate squares or rectangles. Two panels that are available are:

- **Appearance**: With this option we control **Color 1**, **Color 2**, and **Soften**, the last allowing option to blur the transition area. Instead of colors, we may apply other textures.

- **Transforms**: With **Link texture transforms** checked, all transformations made in this panel are also applied to other images used by the material. **Offset** in the **Position** option controls the image origin related to faces and **Rotation** controls rotation. **Scale** controls real-world dimensions, width and height, for the image. **Repeat** controls if the image is repeated to fulfill the faces, that is, **Tile** or not.

Gradient

The **Gradient** texture simulates a smooth transition between two or more colors with the following panels:

- **Appearance**: The **Gradient Type** list includes several gradient shapes, such as **Linear**, **Box**, **Radial**, **Diagonal**, **Sweep**, or **Tartan**. We can pick on the bar to add colors, or drag the small mark out to remove it; a double-click on a mark displays the color box or it can be changed on **Color** (number). The **Interpolation** list includes some options for transition between colors. **Position** controls the relative position for the selected mark, between 0 (left-hand side) and 1 (right-hand side). **Invert Gradient** inverts the color order.

- **Noise**: We can add some irregularities to gradient. **Noise Type** list controls the type, between **Regular**, **Fractal**, or **Turbulence**. **Amount** specifies how much noise is applied and **Size** controls the scale of irregularities. **Phase** controls the animation speed. **Levels**, used for turbulence or fractal noise, controls the number of iterations.

- **Noise Threshold**: These parameters fine-tune **Low** and **High** noise threshold. For instance, with the same value, it is equivalent to not having noise. **Smooth** controls smooth transition level.

- **Transforms**: With **Link texture transforms** checked, all transformations made in this panel are also applied to other images used by the material. **Offset** in the **Positions** option controls the image origin related to faces and **Rotation** controls rotation. **Scale** controls real-world dimensions, width and height, for the image. **Repeat** controls if the image is repeated to fulfill the faces, that is, **Tile** or not.

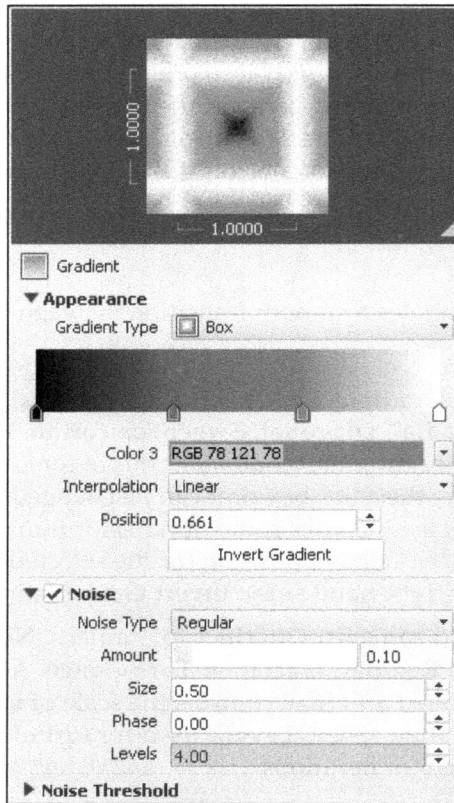

Marble

The **Marble** texture allows you to simulate marbles. This is a 3D texture, that is, it also varies along the object's width. Previous textures are applied by face and don't have transitions between faces. There are two panels:

- **Appearance**: **Stone Color** controls the marble base color and **Vein Color** controls the color of the vein; the texture automatically applies transition colors. **Vein Spacing** controls space between veins with a relative value between 0 and 100; default is **1**. **Vein Width** controls the width of marble veins with a relative value between 0 and 100; the default is **1**.

- **Transforms**: As this is a 3D texture, parameters are slightly different. With **Link texture transforms** checked, all transformations made in this panel are also applied to this texture that is applied to other material characteristics such as relief. **Offset** in the **Positions** option controls the texture origin related to the model corner, in all 3D directions. **Rotation** controls texture rotation in the three directions.

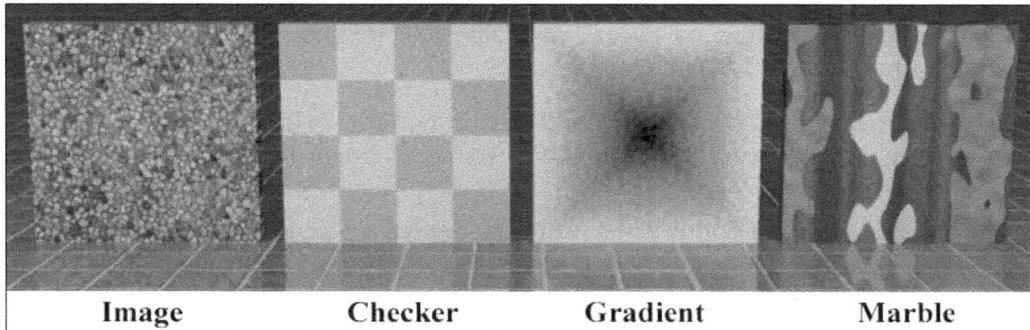

| Image | Checker | Gradient | Marble |

Noise

The **Noise** texture creates an irregular pattern with soft transitions. This is also a 3D texture and is often used to simulate relief and bump or as main texture for irregular materials such as rock or grass. It includes three panels:

- **Appearance**: The **Noise Type** list controls the type, between **Regular**, **Fractal**, or **Turbulence**. **Size** controls the scale of irregularities. **Color 1** and **Color 2** control both colors that are mixed to create the texture.

- **Noise Threshold**: These parameters fine-tune **Low** and **High** noise threshold. **Levels**, used for turbulence or fractal noise, control the number of iterations. **Phase** controls the animation speed.

- **Transforms**: With **Link texture transforms** checked, all transformations made in this panel are also applied to this texture that is applied to other material characteristics such as relief. **Offset** in the **Position** option controls the texture origin related to the model corner, in all 3D directions. **Rotation** controls texture rotation in the three directions.

Speckle

The **Speckle** texture simulates a base texture with small speckles. This is also a 3D texture, used often to simulate relief and bump. It includes two panels:

- **Appearance**: **Color 1** controls speckle color and **Color 2** controls base color. **Size** controls speckle size.

- **Transforms**: With **Link texture transforms** checked, all transformations made in this panel are also applied to this texture that is applied to other material characteristics such as relief. In **Position, Offset** controls the texture origin related to the model corner, in all 3D directions. **Rotation** controls texture rotation around the three directions.

Waves

The **Waves** texture simulates waves. This is also a 3D texture, used mainly to simulate bump. It includes three panels:

- **Appearance**: **Color 1** and **Color 2** control both the colors that are mixed to create the texture. **Distribution** specifies whether it is a **3D** or **2D** distribution of waves.

- **Waves**: **Number** controls the number of wave sets, with a value from 1 to 50; higher values mean less calm water. **Radius** specifies the sphere or circle radius. **Len Min** and **Len Max** define, respectively, the minimum and maximum intervals for wave center; the closer the values, the more regular are the waves. **Amplitude** controls amplitude or frequency of waves. **Phase** shifts the wave. **Random Seed** generates a new random wave distribution within the same parameter.

- **Transforms**: With **Link texture transforms** checked, all transformations made in this panel are also applied to this texture that is applied to other material characteristics such as relief. **Offset** in the **Position** option controls the texture origin related to the model corner, in all 3D directions. **Rotation** controls texture rotation in the three directions.

Wood

The **Wood** texture simulates wood. This is also a 3D texture, mainly applied as base texture. There are two panels:

- **Appearance**: **Color 1** and **Color 2** control both colors that simulate timber bands. **Radial Noise** controls noise in the radial direction with a relative value between 0 and 100; default is **1**. **Axial Noise** controls noise in the axial direction with a relative value between 0 and 100; default is **1**. **Grain Thickness** controls the relative thickness of bands with a value from 0 to 100; default is **0.5**.

- **Transforms**: With **Link texture transforms** checked, all transformations made in this panel are also applied to this texture that is applied to other material characteristics such as relief. **Offset** in the **Position** option controls the texture origin related to the model corner, in all 3D directions. **Rotation** controls texture rotation in the three directions.

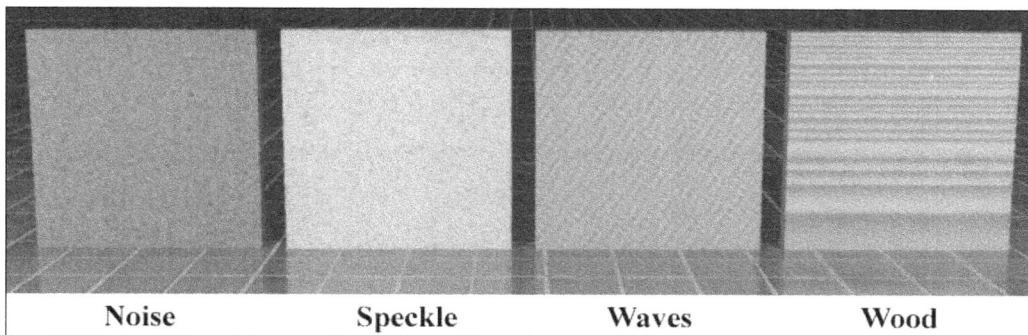

| Noise | Speckle | Waves | Wood |

Tiles

The **Tiles** texture is a powerful 2D texture that can be used to simulate multiple tiles and bricks. It considers two elements, tile and grout (interval between tiles). There are seven panels:

- **Pattern**: The **Type** list displays several presets for tiles pattern, such as **Stack Bond** (all aligned), **Running Bond** (alternated), and including a **Custom** option. **Tile Count** controls how many tiles are **Per Row** and **Per Column**.

- **Tile Appearance**: **Tile Color** allows you to choose a color or applying a texture. **Color Variance** controls a color variance, with a value between 0 and 100; default is **0**. **Fade Variance** controls a fading variance, with a value between 0 and 100; the default is **0.5**. **Randomize** specifies different variances.

- **Grout Appearance**: **Grout Color** allows you to choose a color or applying a texture for grout. **Gap Width** specifies a relative size for grout, with a value between 0 and 100; the default is **0.5**. **Roughness** allows adding roughness to grout, with a value between 0 and 200; the default is **0**.

- **Stacking Layout**: With **Type Custom**, **Line Shift** controls shifting rows, **Random** applies different values when shifting rows, with a value between 0 and 100; the default is **0.5**.

- **Row Modify**: With this checked and by using **Type Custom**, we can modify rows and increase (**Amount** greater than 1) or decrease (**Amount** less than 1) the tile size. **Every** controls the number of rows between changes.

- **Column Modify**: With this checked and with **Type Custom**, you can modify columns and increase (**Amount** greater than 1) or decrease (**Amount** less than 1) the tile size. **Every** controls how many columns between changes.

- **Transforms**: With **Link texture transforms** checked, all transformations made in this panel are also applied to other images used by the material. **Offset** in the **Position** option controls the image origin related to faces and **Rotation** controls its rotation. **Scale** controls real-world dimensions, width and height, for the image. **Repeat** controls if the image is repeated to fulfill the faces, that is, **Tile** or not.

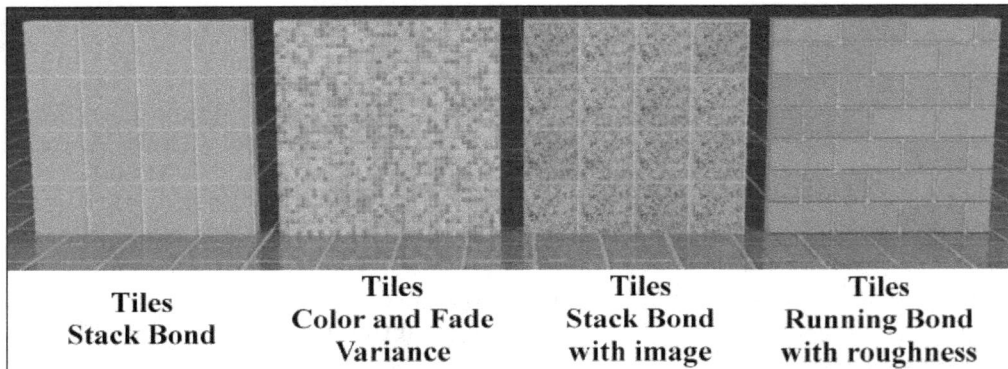

| Tiles
Stack Bond | Tiles
Color and Fade
Variance | Tiles
Stack Bond
with image | Tiles
Running Bond
with roughness |

The best way to analyze texture parameters is by trying. We change parameters and observe the preview image. Sometimes size can be tricky, so we apply the material and the RENDERCROP command (alias RC) to render only part of the scene.

Exercise 11.2

We are going to apply and modify materials on the model that we used for lighting:

1. Open the drawing A3D_11_02.DWG. The lighting is already set up to photometric, with sunlight and skylight.

2. This model needs materials for house walls, stone, roof, terrace, floor, terrain, windows, and door.

3. With MATBROWSEROPEN command (alias MAT) load the materials, such as **Cool White** (the **Wall Paint** library), **Rubble - River Rock** (the **Masonry/Stone** library), **Spanish Tile - Red 2** (the **Roofing** library), **3in Square - Terra Cotta** (the **Ceramic** library), **Diamond - Red** (the **Flooring** library), **Grass - Thick** (the **Sitework** library), **Clear** (the **Glass** library), and **Mahogany - Solid Stained Dark Medium Gloss** (the **Wood** library).

4. With the MATERIALATTACH command, assign materials to layers as displayed in the next screenshot:

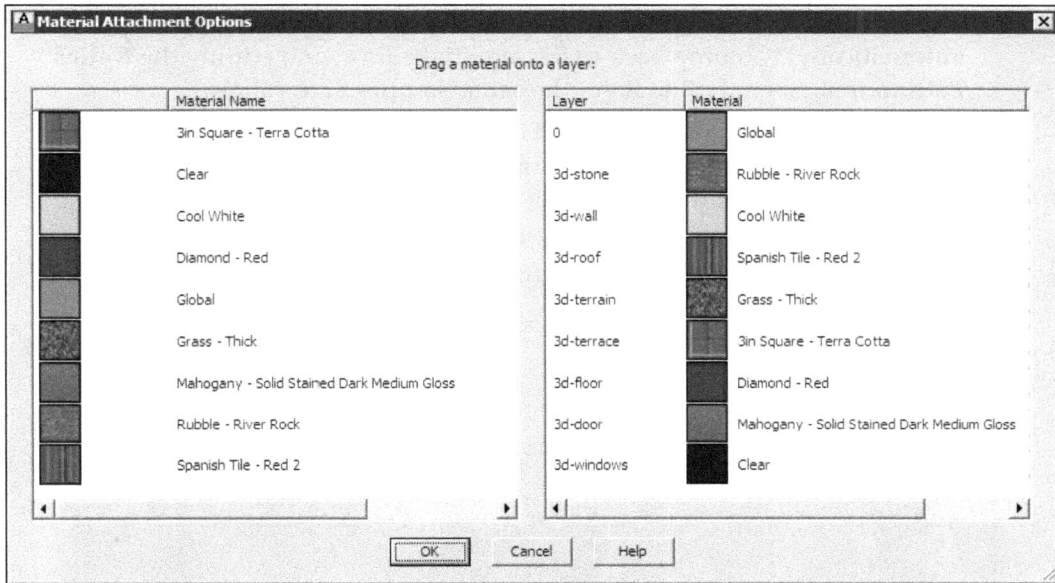

5. Rendering the scene, the materials are applied, but we need to adjust material scales.

6. The white wall material, **Cool White**, was created from the **Wall Paint** template and need not be changed.

7. The tower rock material, **Rubble - River Rock**, needs to be modified. On the **Material Browser** palette, double-click on it and, on the **Materials Editor** palette, edit the image. On the **Texture Editor** palette, activate **Link texture transforms** and modify **Sample Size** to **1.5** width (height adjusts automatically). Coming back to the material editor, and editing the **Relief Pattern** image, verify that it got the same **Sample Size** values.

8. The tower roof material, **Spanish Tile - Red 2**, also needs to be adjusted. Editing the main image, activate **Link texture transforms** and modify **Sample Size** to **1** width.

9. Adjusting the terrace material, **3in Square - Terra Cotta**, on the **Texture Editor**, activate **Link texture transforms** and modify **Sample Size** to **0.5**.

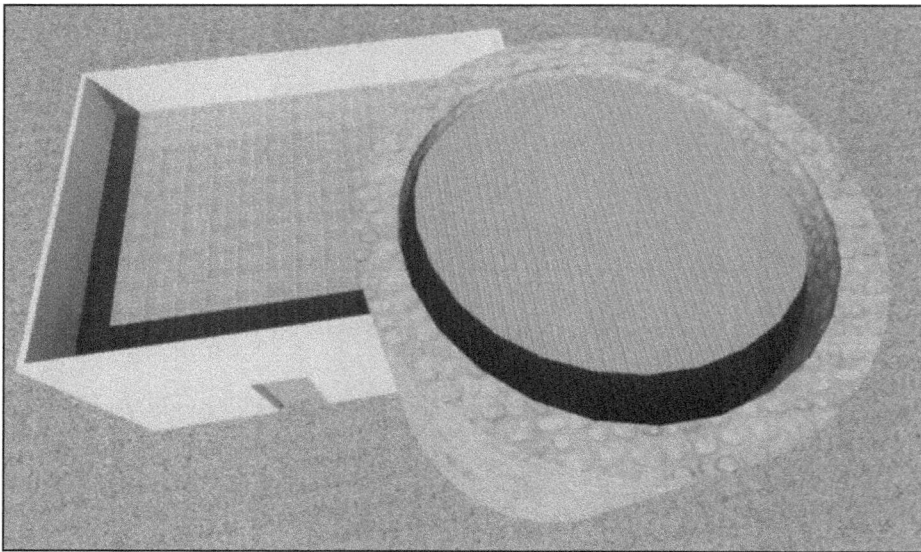

10. To verify materials inside the house, hide several objects, or activate one of the cameras that are inside. In the last case, you must turn on all lights. We are going to apply the last solution. Select the view from Camera1 (with the VIEW command or middle viewport control).

11. With LIGHTLIST, select all five lights by right-clicking, select **Properties**, and turn on all lights with only one operation. With SUNPROPERTIES you may need to activate **Sky Background and Illumination**.

12. The inner floor material, **Diamond - Red**, does not have the correct scale and is too reflective. Modify textures **Sample Size** to **1** and, under the **Reflectivity** panel we decrease **Direct** to **20**.

13. With the help of **Previous** in the **ZOOM** options, come back to the exterior view.

14. Adjust the sample size of the terrain material, **Grass - Thick**, to **3** units.

15. The windows material, **Clear**, does not need any modifications.

16. Finally, the door material, **Mahogany - Solid Stained Dark Medium Gloss**, that was created from the **Wood** template; change its texture applied in **Image**, decreasing **Brightness** to **50** and specifying **Sample Size** to **0.5**.

17. Exterior renders should now display good quality materials.

18. Save the drawing with the name A3D_11_02final.DWG.

Mapping coordinates

After selecting and modifying materials, when we apply these to objects, sometimes we need to change something on the objects. This is called mapping coordinates and represent the way materials are painted on faces and surfaces. To distinguish mapping coordinates from modeling coordinates, letters U, V, and W are used, U and V being face coordinates and W the normal to face. The next command controls mapping coordinates.

The MATERIALMAP command

The MATERIALMAP command (no alias) allows you to modify mapping coordinates for selected objects or selected faces. Mapping coordinates can be adjusted to geometry faces, such as **Planar**, **Box**, **Spherical**, or **Cylindrical** and then, moved or rotated. By applying the command from the **Render** toolbar or ribbon, we can specify the geometry directly.

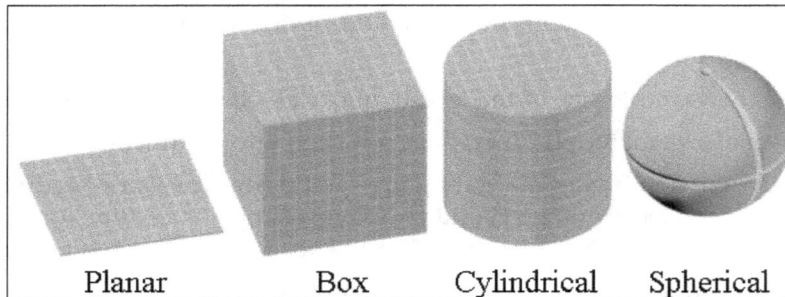

| Planar | Box | Cylindrical | Spherical |

First, the command prompts for the geometry mode:

```
Command: MATERIALMAP
Select an option [Box/Planar/Spherical/Cylindrical/copY mapping to/
Reset mapping] <Box>: Option
```

Then, we must select objects. The command accepts solids, surfaces, meshes, regions, and 2D objects with thickness. We can also select object faces by pressing *Ctrl*:

```
Select faces or objects: Selection
```

We accept the mapping or apply an option. The prompt repeats until acceptance:

```
Accept the mapping or [Move/Rotate/reseT/sWitch mapping mode]: Enter
```

The options for this command are:

- **Box/Planar/Spherical/Cylindrical**: We specify the most suitable geometry to the object

- **Copy mapping to**: This allows you to copy mapping coordinates from an object and pass them to the objects to be selected

- **Reset mapping**: This option restores the original mapping coordinates

- **Move**: This option displays the 3DMOVE gizmo and grips, allowing you to move the material on the object by specifying two points

- **Rotate**: This option displays the 3DROTATE gizmo, allowing you to rotate the material on the object by specifying a rotation axis and a point to indicate rotation angle

- **Reset**: This option restores the original mapping coordinates

- **Switch mapping mode**: This displays the initial prompt, allowing you to select a different geometry

> The command does not allow you to specify direct values for moving or rotating coordinates, which is a severe limitation.
>
> Sometimes, AutoCAD is not able to display correct mapping coordinates. In these situations, applying this command should correct the problem.

Exercise 11.3

Applying materials to a simple roof, we will demonstrate how to adjust mapping coordinates and how to assign a different material to faces:

1. Open the drawing A3D_11_03.DWG. The lighting is already set up to photometric, with sunlight and skylight.

2. With MATBROWSEROPEN, load the **Spanish Tile - Red 2** material (the **Roofing** library). Also create a new **Wall Paint** material.

3. Editing the first material, access the **Generic** image palette, activate **Link texture transforms,** and assign **Sample Size** the value **2** in width.

4. Editing the second material, just modify its name to White Inner walls.

5. For this simple case, instead of assigning materials by layer, drag the material from the **Materials Browser** palette to the roof. We can immediately see that the material is not applied correctly.

6. We have six different roof planes, but two are parallel. Apply the MATERIALMAP command, specify **Planar**, and select both front faces with the *Ctrl* key pressed. The rotate gizmo and a yellow rectangle appear. Now select the blue circle on the rotate gizmo and rotate the rectangle by eye (sorry, there is no other way) until the rectangle aligns with the lower edge. Click to specify this rotation and press *Enter* to end the command.

7. Repeat this procedure until all faces have the corrected mapping coordinates.

8. Now apply the white paint material to all inner and bottom faces. With the **Materials Browser** palette open, press the *Ctrl* key and select all inner faces and the bottom face. By right-clicking on the material select **Assign to Selection**.

9. Save the drawing with the name `A3D_11_03final.DWG`.

Effects

AutoCAD considers, as effects, only scene background and fog or depth cue.

Backgrounds

As scene background, we may automatically apply the sky background (if working with photometric lighting and perspective) or one of the three options, namely solid color, color gradient, or image. The `VIEW` command, presented in *Chapter 2*, *Visualizing 3D Models* allows you to specify a background that is associated with the view. Another way is the `BACKGROUND` command.

The BACKGROUND command

The BACKGROUND command (no alias) allows you to specify a background for the current view and next visualizations. The **2D wireframe** visual style does not allow background visualization in the viewport. This command is not available in menus, toolbars, or ribbon and neither is it documented as a command in **Help**.

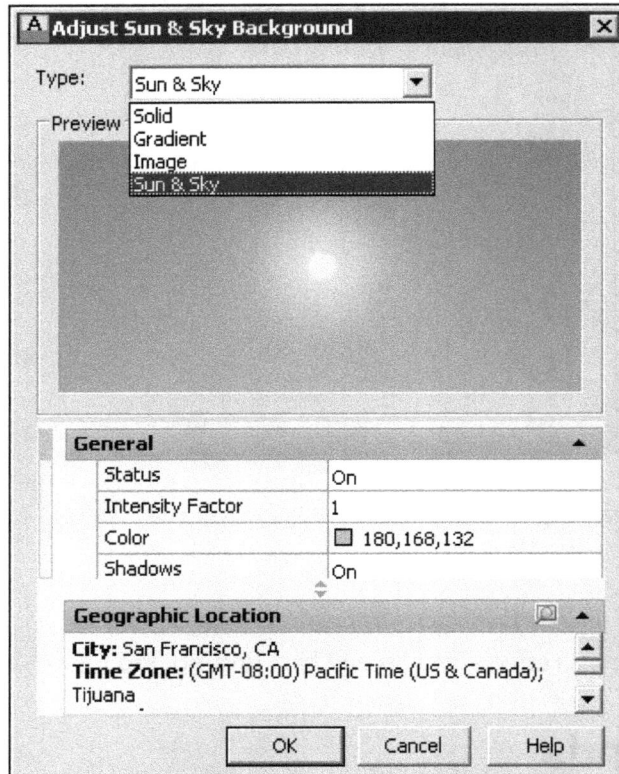

The command displays a dialog box with a preview area. The **Type** list controls which background is applied:

- **Sun & Sky**: The sky color and even the Sun disc visualization are calculated based on Sun azimuth and altitude. This type is only available with photometric lighting and perspective.

- **Solid**: We select a color for a uniform background. Picking the color slot displays the AutoCAD color dialog box.

- **Gradient**: We apply a gradient background specifying two or three colors (with **Three color** checked). We select each color and may specify a rotation value.

- **Image**: In **Browse** we select an image to be applied as the background. The **Adjust Image** button displays a new dialog box to adjust the chosen image to the background resolution (normally the render resolution). In this box, the **Image position** list includes the **Center**, **Stretch**, and **Tile** options. **Center** places the image at the render and viewport centers, but if resolutions are different, it can be cut or left with blank margins. **Stretch** stretches the image to the render resolution. **Tile** repeats the image until background resolution is fulfilled. Both horizontal and vertical sliders, not available to **Stretch**, can adjust image position and image scale. With **Maintain aspect ratio when scaling** activated, both scales are locked to each other.

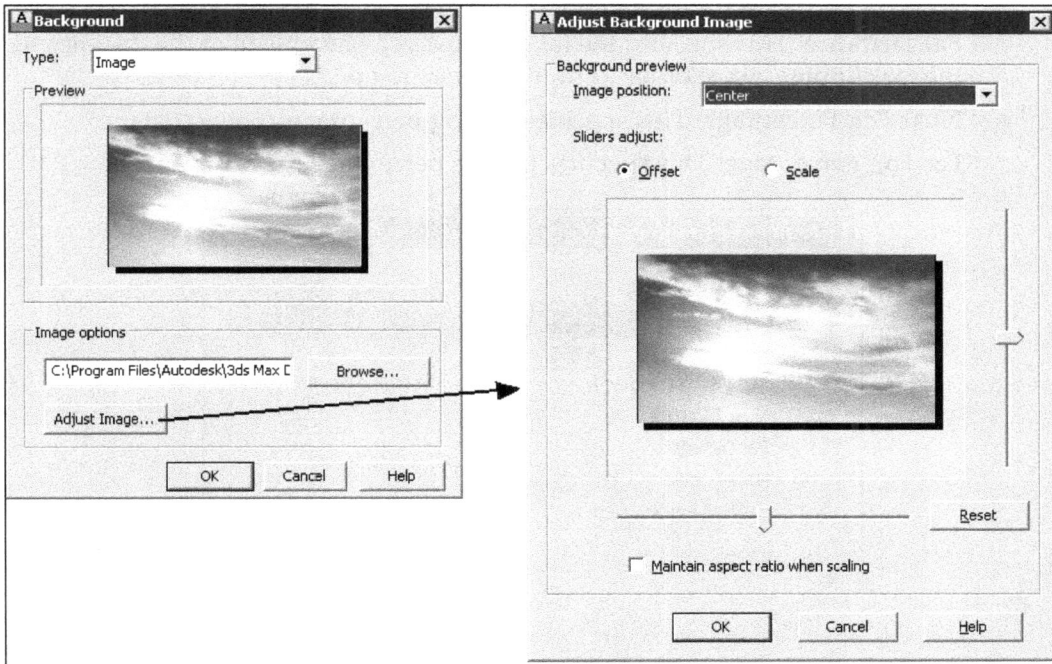

Fog and depth cue

A command is available to define fog or depth cue.

The RENDERENVIRONMENT command

The RENDERENVIRONMENT command (alias FOG, which was the old name) allows you to define fog to exterior scenes and also depth cue, that is, loss of visibility with distance. The command displays a dialog box with the following parameters:

- **Enable Fog**: With this option, we activate either fog or depth cue
- **Color**: We select white or light gray, and black for depth cue
- **Fog Background**: If this is on, fog is also applied to the background
- **Near Distance**: This specifies the near distance, as a percentage of the distance to the far clipping plane or background, where fog begins
- **Far Distance**: This specifies the far distance as a percentage of the distance to the far clipping plane or background, where fog ends
- **Near Fog Percentage**: This specifies the fog percentage at near distance
- **Far Fog Percentage**: This specifies the fog percentage at far distance

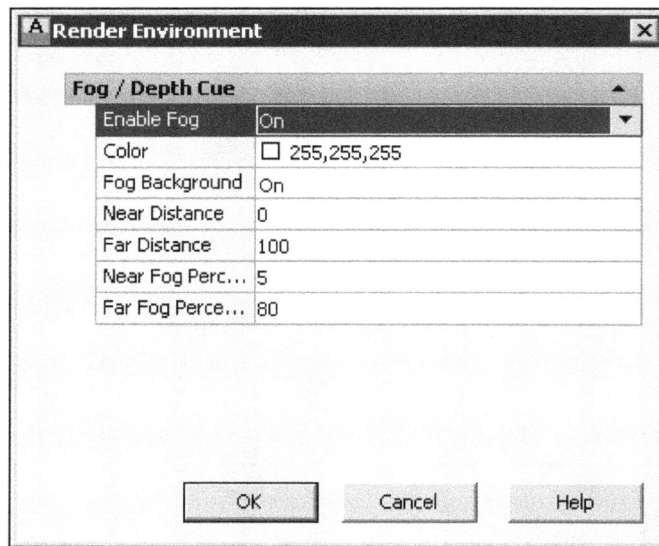

> Autodesk needs to update this command, as it is very tricky. Sometimes we change values and nothing happens. One trick that does not work every time is to change camera view.

Summary

This chapter includes all commands and concepts related to materials and environment. We started with an introduction to materials and textures followed by an explanation of the Autodesk Material Library, which includes more than 700 materials almost ready to assign, and where to find more than 1,000 textures available to create our own materials. The MATBROWSEROPEN command displays a palette that allows loading and managing the scene materials, also assigning these to objects.

The advised way to assign materials is by layer, which can be achieved with the MATERIALATTACH command.

From the MATBROWSEROPEN palette we may create materials based on several material templates, such as **Generic**, **Ceramic**, **Concrete**, and others. To define or modify materials, we have the MATEDITOROPEN command, also available from the previous palette by double-clicking on a material slot. To several material parameters, we can apply images and procedural textures/maps, such as **Checker**, **Tiles**, and **Noise**. To edit images or textures, such as defining scales, we have the **Texture Editor** palette.

Sometimes, material textures are not well placed on faces, so we need to adjust mapping coordinates, with the MATERIALMAP command.

Two of the effects available are, background that can be defined with the VIEW command or the BACKGROUND command; and fog or depth cue that is defined with the RENDERENVIRONMENT command.

In the next chapter we will present 3D meshes and surfaces.

12
Meshes and Surfaces

In *Chapter 4*, *Creating Solids and Surfaces from 2D*, we introduced how to create surfaces by applying EXTRUDE and other commands to linear objects. This chapter is about all types of surfaces and meshes.

The topics covered in this chapter are:

- Identifying types of surfaces and their applications
- How to create and edit procedural surfaces
- How to create and edit NURBS surfaces
- How to create and edit meshes
- How to create meshes based on 3D faces
- Analyzing surfaces curvature

Surfaces and meshes are 3D objects with zero thickness. Most of the time, when modeling in 3D, we apply solids, which give us much better precision; since Version 2010, surfaces and meshes are also available, and can be used for complex and conceptual models. These objects are divided into the following categories:

- **Procedural surfaces**: These are surfaces created from linear objects, such as EXTRUDE, REVOLVE, or SWEEP, as we have seen in *Chapter 4*, *Creating Solids and Surfaces from 2D*. These surfaces could or could not maintain association with original linear objects. We will present several other commands for creating procedural surfaces.

- **NURBS surfaces**: **Non-Uniform Rational B-Splines** (**NURBS**) surfaces are based on splines and are mainly modified by control vertices. This type of surfaces is mainly applied in the aeronautical, automotive, mould, or any industry with free-form shapes.

- **Meshes**: These objects, also called tessellation meshes, are not really surfaces but allow creating smooth 3D objects.

- **Polyface meshes**: These are the oldest AutoCAD 3D objects, long before solids, and are based on 3D faces.

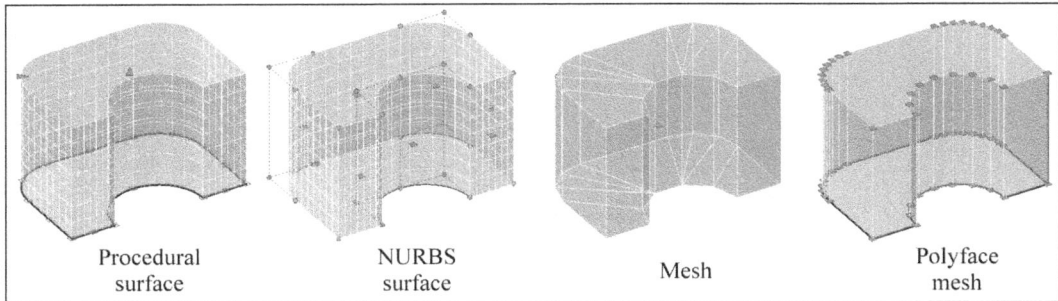

| Procedural surface | NURBS surface | Mesh | Polyface mesh |

Procedural surfaces

Procedural surfaces can be created from linear objects, with some commands presented in *Chapter 4, Creating Solids and Surfaces from 2D*, such as EXTRUDE or SWEEP, also with PLANESURF, presented in *Chapter 5, 3D Primitives and Conversions*, and with SURFNETWORK, presented next section. These surfaces can also be obtained from other surfaces, applying known operations such as blend, fillet, or offset. Included are also two editing commands and a command to create solids from surfaces.

The SURFACEMODELINGMODE variable must have value 0. With value 1, NURBS surfaces are created instead when applying next commands. The SURFU and SURFV variables control the surface isolines density in the first and second directions and can be modified any time with the PROPERTIES palette.

Access to commands

Most procedural and NURBS surface commands are on the ribbon, **Surface** panel, on the **Draw/Modeling** and **Modify/Surface Editing** menu bars, and on the **Surface Creation** and **Surface Editing** toolbars.

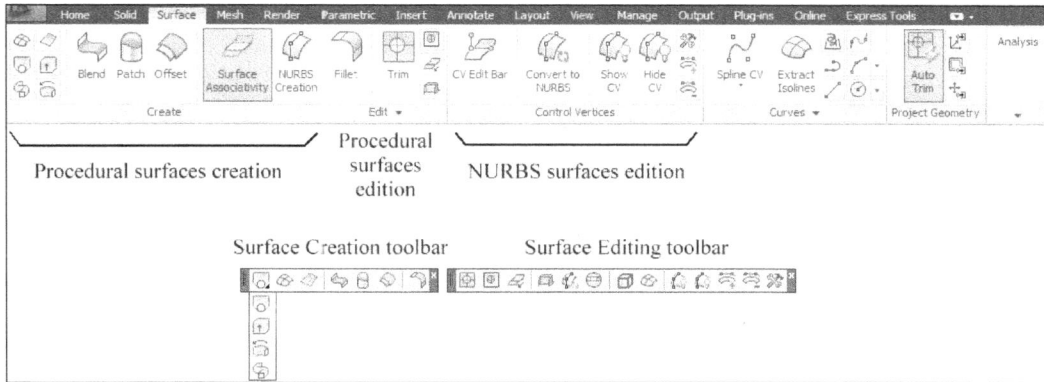

Procedural surfaces creation

Procedural surfaces edition

NURBS surfaces edition

Surface Creation toolbar

Surface Editing toolbar

Creating surfaces from linear objects and by conversion

In *Chapter 4, Creating Solids and Surfaces from 2D* we already introduced several commands for creating procedural surfaces. Let's summarize them:

- EXTRUDE (alias EXT): This command creates surfaces by applying an extrusion height to open linear objects or closed linear objects if the **MOde** option is set to **SUrface**.

- REVOLVE (alias REV): This command creates surfaces by rotating open linear objects, or closed linear objects if the **MOde** option is set to **SUrface**, around an axis.

- SWEEP: This command creates a surface by extruding an open section, or closed section if the **MOde** option is set to **SUrface**, along a path.

- LOFT: This command creates surfaces from the selection of two or more open linear cross sections or closed linear cross sections if the **MOde** option is set to **SUrface**.

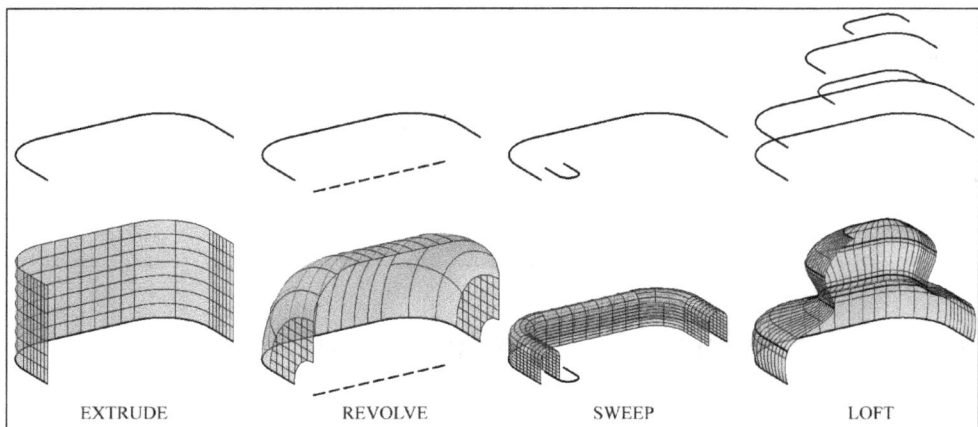

| EXTRUDE | REVOLVE | SWEEP | LOFT |

When applying these commands from the **Surface Creation** toolbar or ribbon, **Surface** panel, the **MOde** option is already set to **SUrface**. These commands also generate surfaces if applied to edges of solids or surfaces (selected by pressing the *Ctrl* key).

The SURFACEASSOCIATIVITY variable controls if surfaces maintain an association to original linear objects. If the value is 0, association is not maintained; if the value is 1 (default) association is maintained. This variable can be directly manipulated on the ribbon.

Two other commands, presented in *Chapter 5, 3D Primitives and Conversions*, also allow creating surfaces:

- PLANESURF: This command creates planar surfaces defined by two opposite corners or delimited by one or more existing objects. We can select lines, arcs, circles, ellipses, 2D polylines, 3D planar polylines, planar splines, regions, and 3D faces. For each closed object or set of objects that form a closed boundary, a surface is created.

- CONVTOSURFACE: This command converts the following objects to surfaces, namely 2D open polylines with nonzero thickness, lines and arcs with nonzero thickness, regions, planar 3D faces, 2D solid objects (SOLID), 3D solids and 3D meshes.

The SURFNETWORK command

The SURFNETWORK command (no alias) allows you to create a surface from two sets of open linear objects that specify a boundary and net in two directions. The command starts by prompting the objects in the first direction:

```
Command: SURFNETWORK
Select curves or surface edges in first direction: Selection
```

We press *Enter* to finish the first selection:

```
Select curves or surface edges in first direction: Enter
```

We select the second set of objects:

```
Select curves or surface edges in second direction: Selection
```

When pressing *Enter*, the surface is created:

```
Select curves or surface edges in second direction: Enter
```

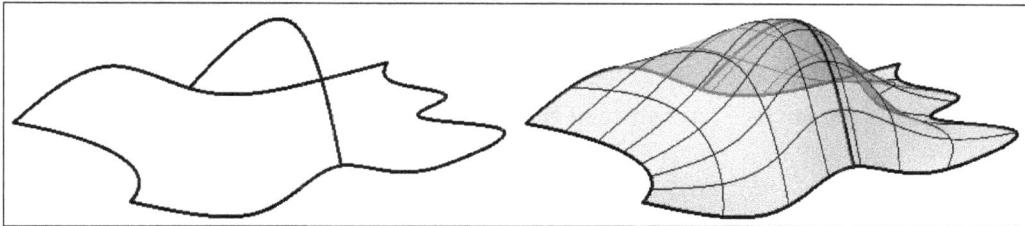

> If the SURFACEASSOCIATIVITY variable is 1, by modifying linear objects, the surface immediately adjusts.
>
> In the selection sets we can include surfaces and solids edges.

Creating surfaces from other surfaces

Several commands are presented that allow you to create surfaces from existing surfaces. Bulge and continuity are two options frequently used when creating or adjusting surfaces.

Continuity and bulge

The two important surface properties are continuity and bulge. These properties can be controlled when creating the surface (with options or grips) or later, with grips or the PROPERTIES palette.

Continuity defines the transition between surfaces and AutoCAD allows three possibilities:

- G0: This is just position, coincident edges between the two surfaces, with crease

- G1: This is position and tangency; there is a smooth transition (no crease) between surfaces but can have sudden changes in curvature

- G2: This is position, tangency, and curvature; the two surfaces share the same curvature or a similar radius at the transition

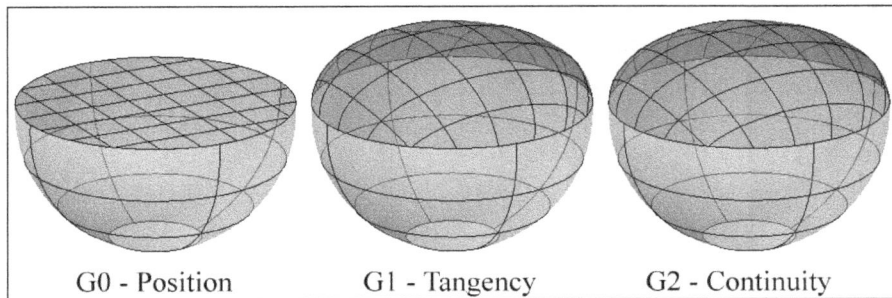

| G0 - Position | G1 - Tangency | G2 - Continuity |

Bulge controls the surface curvature at transition edges, with a value between 0 and 1.

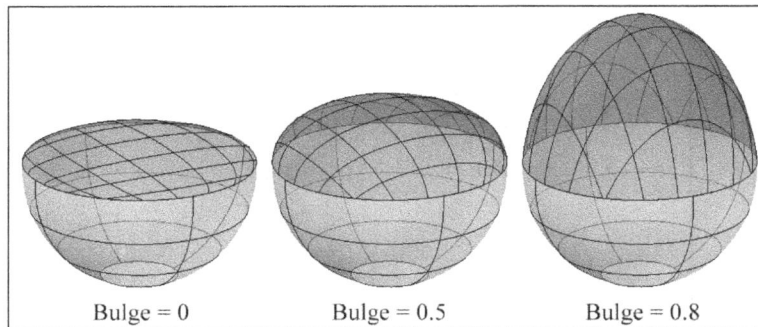

| Bulge = 0 | Bulge = 0.5 | Bulge = 0.8 |

The SURFBLEND command

The SURFBLEND command (no alias) creates a surface connecting two existing surfaces. The command just prompts for the selection of two sets of edges from the surfaces to connect and allows for controlling continuity and bulge:

```
Command:  SURFBLEND
```

Information is displayed with default continuity and bulge values:

```
Continuity = G1 - tangent, bulge magnitude = 0.5
```

The command prompts for edges selection on the first surface, finished by pressing *Enter*. The **CHain** option allows for selecting all connected edges by selecting only one:

```
Select first surface edges to blend or [CHain]: Selection
Select first surface edges to blend or [CHain]: Enter
```

The command prompts for edges selection on the second surface, finished by pressing *Enter*:

```
Select second surface edges to blend or [CHain]: Selection
Select second surface edges to blend or [CHain]: Enter
```

We press *Enter* to accept the surface or apply options to control continuity or bulge values:

```
Press Enter to accept the blend surface or [CONtinuity/Bulge
magnitude]: Enter
```

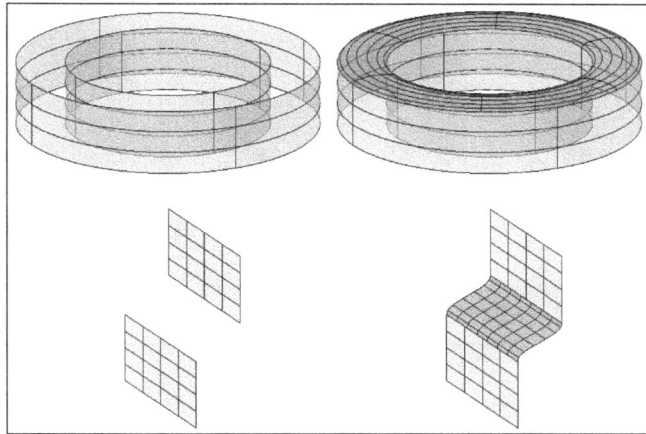

The SURFPATCH command

The SURFPATCH command (no alias) creates a surface that closes a hole or an opening. By default, the command prompts for the selection of edges that limits the opening:

```
Command: SURFPATCH
```

Information is displayed with default continuity and bulge values:

```
Continuity = G0 - position, bulge magnitude = 0.5
```

The command prompts for edges selection on the first surface, finished by pressing *Enter*. The **CHain** option allows for selecting all connected edges by selecting only one, and the **CUrves** option allows selecting linear entities:

```
Select surface edges to patch or [CHain/CUrves] <CUrves>: Selection
Select surface edges to patch or [CHain/CUrves] <CUrves>: Enter
```

We press *Enter* to accept the surface or apply options to control continuity or bulge values. The **Guides** option allows selecting guided lines that mould the surface:

```
Press Enter to accept the patch surface or [CONtinuity/Bulge
magnitude/Guides]: Enter
```

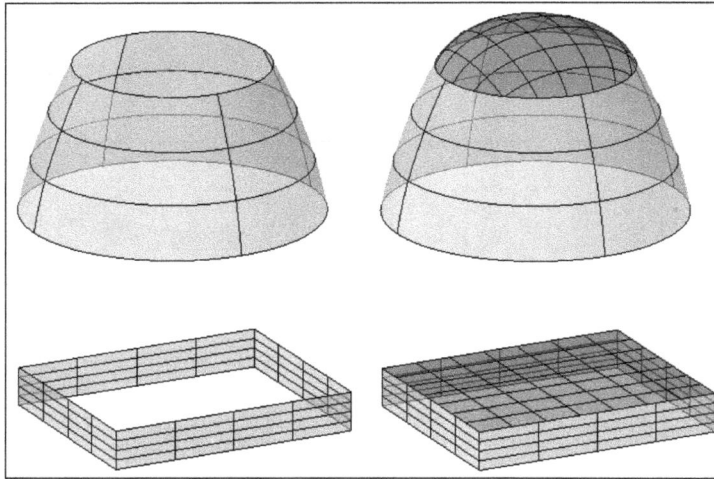

The SURFEXTEND command

The SURFEXTEND command (no alias) creates a surface by extending existing surfaces from selected edges. By default, the command prompts for edges selection and distance:

```
Command: SURFEXTEND
```

Information is displayed with the default values for the **Modes** option:

```
Modes = Extend,  Creation = Append
```

We select one or more edges from which surfaces will extend:

```
Select surface edges to extend: Selection
Select surface edges to extend: Enter
```

We specify the extension distance. The **Expression** option allows entering a mathematical expression. The **Modes** option includes the possibility of stretching the surface (**Stretch**), instead of extending (**Extend**), and merging into the existing surface (**Merge**), instead of creating a new one (**Append**):

```
Specify extend distance [Expression/Modes]: Value
```

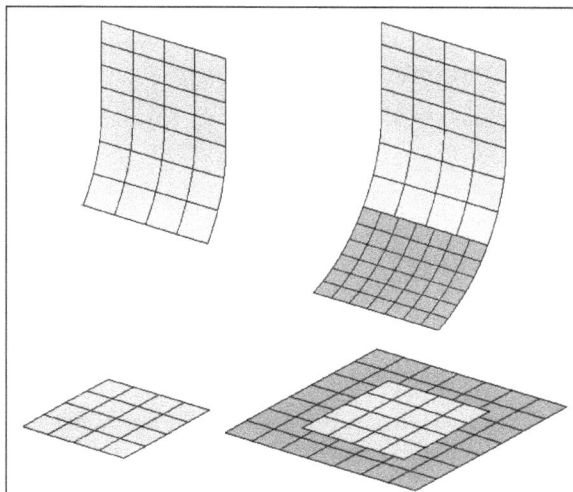

The SURFOFFSET command

The SURFOFFSET command (no alias) creates surfaces parallel to the selected surfaces.

By default, we select surfaces and specify a distance:

```
Command: SURFOFFSET
```

Information with the default value for the **Connect** option:

```
Connect adjacent edges = No
```

The command prompts for the surfaces' selection:

```
Select surfaces or regions to offset: Selection
Select surfaces or regions to offset: Enter
```

Small arrows, normal to surfaces, are displayed indicating the offset direction. By default, we only need to specify a distance. The **Flip direction** option allows you to invert the direction. The **Both sides** option applies offset in both directions. The **Solid** option, creates a solid corresponding to the volume between the surfaces, similar to the THICKEN command application. The **Connect** option also connects offset surfaces if original surfaces are connected:

```
Specify offset distance or [Flip direction/Both sides/Solid/Connect/
Expression] <0.0000>: Value
```

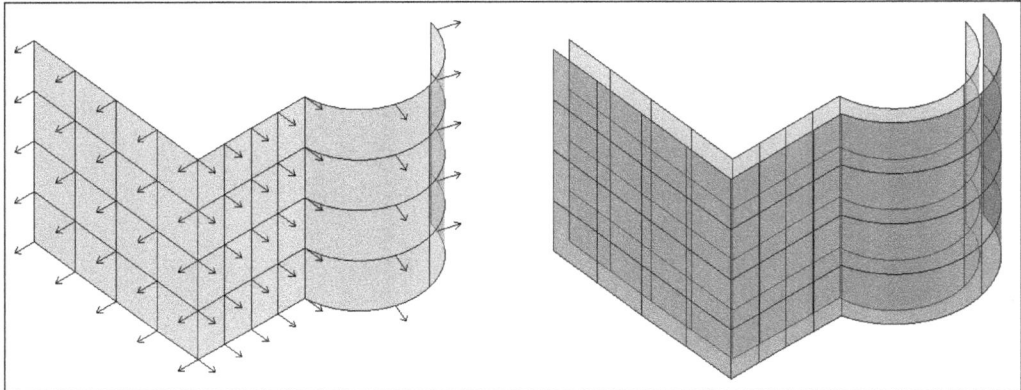

> Unlike other surface creation commands, surfaces created with this command maintain layers and U and V isolines of the original surfaces, instead of the current layer.

The SURFFILLET command

The SURFFILLET command (no alias) creates a fillet surface between two existing surfaces. The fillet radius is adjusted by grip or option.

```
Command: SURFFILLET
```

Information with default values for command options:

```
Radius = 6.0000, Trim Surface = yes
```

The command prompts for the selection of the first surface and, then, the second surface:

```
Select first surface or region to fillet or [Radius/Trim surface]:
Selection
Select second surface or region to fillet or [Radius/Trim surface]:
Selection
```

A preview surface is displayed with a grip for controlling fillet radius. The **Radius** option allows defining a value for radius. The **Trim surfaces** option trims and extends, or not, original surfaces to meet the fillet surface.

```
Press Enter to accept the fillet surface or [Radius/Trim surfaces]:
Enter
```

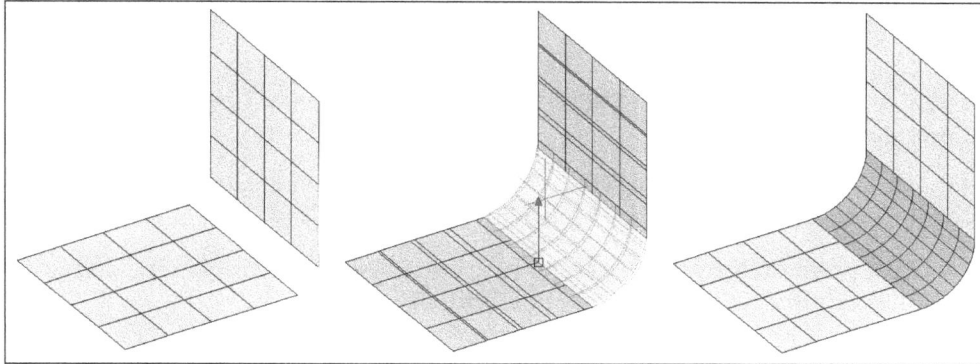

Editing surfaces

Frequently we need to trim surfaces. Related to that, we presented the PROJECTGEOMETRY command in *Chapter 7, Editing Solids and Surfaces*. If the SURFACEAUTOTRIM variable has value 1, that command, besides projecting linear objects, also cuts surfaces and solids. Now we refer two commands specific to that operation.

The SURFTRIM command

The SURFTRIM command (no alias) cuts surfaces with other surfaces, linear objects, or regions. By default, the command prompts for surfaces or regions to trim, the trimming geometry, and the side that will be trimmed. Trimmed surfaces can be restored with the next command.

```
Command: SURFTRIM
```

Information with default values for command options:

```
Extend surfaces = Yes, Projection = Automatic
```

The command prompts for surfaces or regions to be trimmed. The **Extend** option controls if intersecting objects can be extended for boundaries definition, or must really intersect. The **PROjection** direction defines how trimming geometry is projected onto surfaces to trim, may be **Automatic**, **View**, **UCS**, or **None**:

```
Select surfaces or regions to trim or [Extend/PROjection direction]:
Selection
Select surfaces or regions to trim or [Extend/PROjection direction]:
Enter
```

The command prompts for objects (surfaces, regions, or linear objects) that define trimming boundaries:

```
Select cutting curves, surfaces or regions: Selection
Select cutting curves, surfaces or regions: Enter
```

We specify a point on the area or side to be trimmed. The prompt is repeated until the *Enter* key is pressed. The **Undo** option undoes last trim:

```
Select area to trim [Undo]: Point
Select area to trim [Undo]: Enter
```

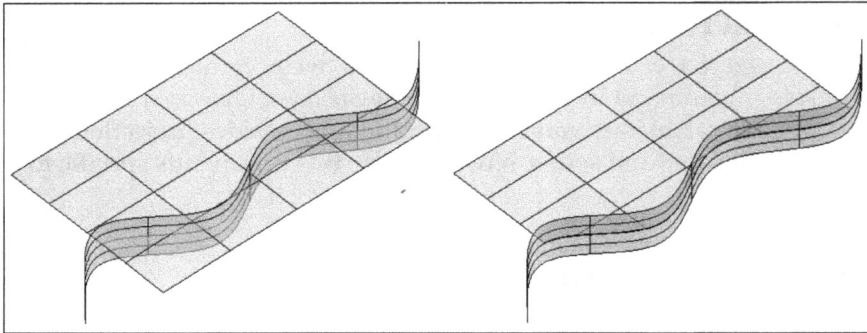

The SURFUNTRIM command

The SURFUNTRIM command (no alias) restores parts of trimmed surfaces. It only prompts for the selection of edges resulting from trimmed operations. The **SURface** option allows restoring all trimmed parts on the selected surfaces:

```
Command: SURFUNTRIM
Select edges on surface to un-trim or [SURface]: Selection
Select edges on surface to un-trim or [SURface]: Enter
```

Exercise 12.1

From a 2D simple drawing, we are going to model a building with surfaces by following these steps:

1. Open the drawing A3D_12_01.DWG. The layers are already created.

2. Active the 3D-Walls layer. Extrude the hexagon by 6 units, as surface (with **MOde** option set to **Surface**, or applying the **Surface Creation** toolbar or **Surface** tab). Extrude the two lines by 3 units.

3. Create an inner surface, apply the SURFOFFSET command, select the hexagonal surface, change the **Flip direction** option, and specify an offset distance of 12 units.

4. We apply SURFFILLET with a radius of 4 units to the corner between the two straight surfaces.

5. Activating the **X-Ray** visual style, the model should look like the next diagram:

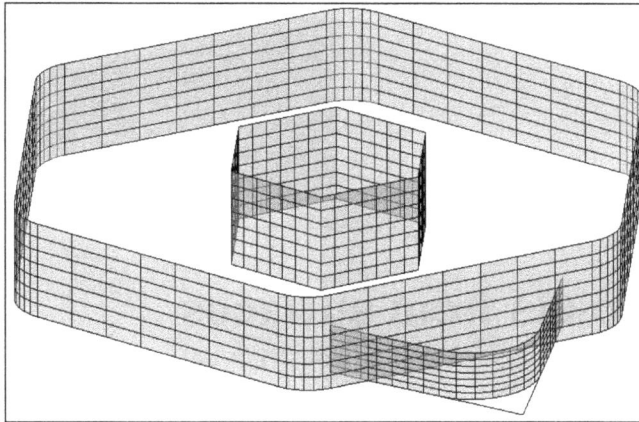

6. Activate the 3D-Roofs layer. Applying the SURFBLEND command, specify the **CHain** option and select an edge on the outer surface. Then, again with the **CHain** option, select an edge on the inner surface and press *Enter* to accept the blended surface.

7. To cut the parts of the straight surfaces that are inside the outer surface, the SURFTRIM command should be applied. Select both straight surfaces, press *Enter*, and select the outer surface as cutting surface. Finally, mark on the parts of the straight surfaces that will be cut.

8. To create a roof on the entrance of the building, we need an extra surface. Draw a line between top endpoints of the straight surfaces and extrude it with -1.

9. Now, apply the SURFPATCH command, selecting all edges that form the closed loop, and confirm a flat patch surface.

10. Now delete the small extruded surface.

11. Finally, trim the outer surface with the SURFTRIM command, selecting the outer surface, as cutting surfaces we select the straight surfaces and the entrance roof, and specify the area to be trimmed.

12. Freeze the 2D layer.

13. Save the drawing with the name A3D_12_01final.DWG.

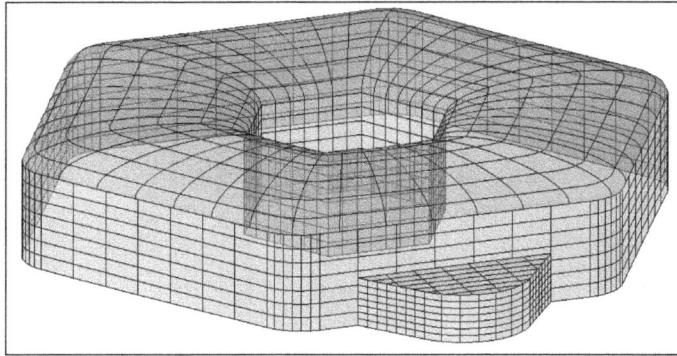

Creating solids from surfaces

The next command does not create or edit surfaces but is closely associated.

The SURFSCULPT command

The SURFSCULPT command (no alias) creates a solid from a closed volume, delimited by surfaces, regions, meshes, or existing solids. Surfaces and regions used in this creation are deleted and solids are united to the resulting solid. Surfaces must be G0, that is, with no tangencies between them:

```
Command: SURFSCULPT
```

Information about the SMOOTHMESHCONVERT variable (used if meshes are included):

```
Mesh conversion set to: Smooth and optimized.
```

The command only prompts for surfaces, solids, regions, and meshes that completely enclose a volume:

```
Select surfaces or solids to sculpt into a solid: Selection
Select surfaces or solids to sculpt into a solid: Enter
```

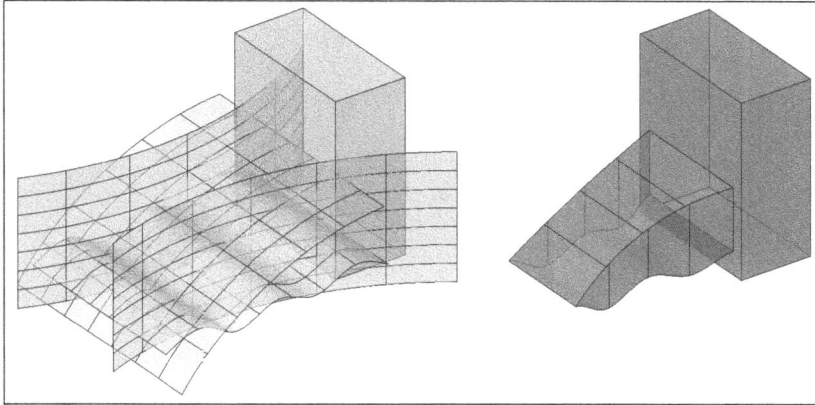

> The SMOOTHMESHCONVERT variable controls the face smoothness when converting meshes (presented later in this chapter) to solids or surfaces.

Exercise 12.2

Let's apply this command to create a solid from the model created in the *Exercise 12.1* section:

1. Open the drawing A3D_12_02.DWG.

2. To have an enclosed volume, the model must be capped below. So we have to create PLANESURF involving all model base.

3. Now, it's only applying the SURFSCULPT command and selecting all objects. The solid is created and all surfaces are deleted.

4.   Save the drawing with the name A3D_12_02final.DWG.

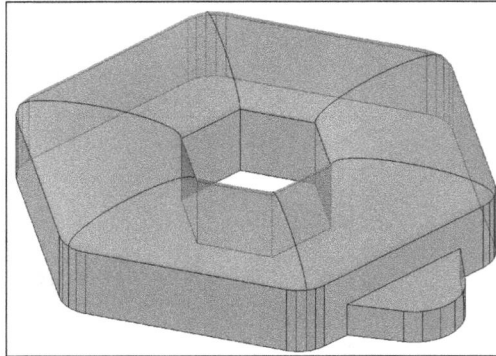

From this model, we could continue our model by applying a shell (SOLIDEDIT command) and making openings for doors and windows.

NURBS surfaces

NURBS surfaces are based on splines and are mainly used for free-form surfaces such as those used in the automotive industry. There are two processes for creating NURBS surfaces and several commands to edit them, which are presented next.

Creating NURBS surfaces

We can create NURBS surfaces by previously specifying a value 1 for the SURFACEMODELINGMODE system variable and applying surface creation commands, or by converting existing objects with the CONVTONURBS command.

The SURFACEMODELINGMODE variable

When creating surfaces, the SURFACEMODELINGMODE variable controls if NURBS or procedural surfaces are created:

-   0: With this value, a procedural surface is created when creating surfaces.
-   1: With this value, a NURBS surface is created when creating surfaces. It is applied to all surface creation commands, with the exception of PLANESURF.

This variable is not saved and by default is set to 0.

The CONVTONURBS command

The CONVTONURBS command (no alias) allows you to convert procedural surfaces and solids to NURBS surfaces. It only prompts for objects selection:

```
Command: CONVTONURBS
Select objects to convert to: Selection
```

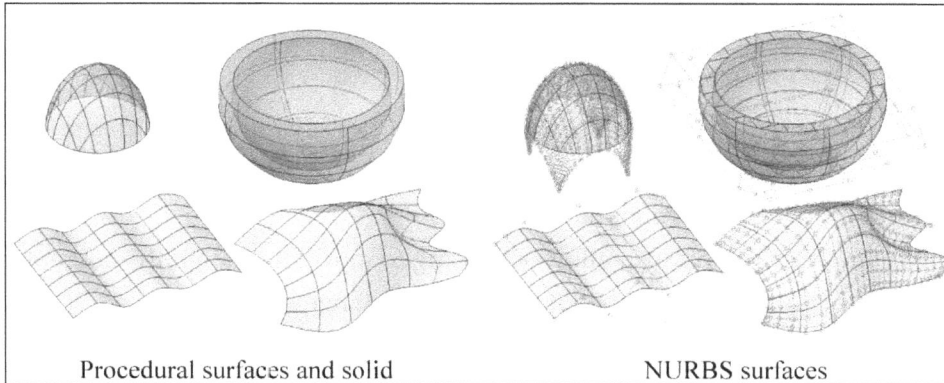

Procedural surfaces and solid   NURBS surfaces

> NURBS surfaces cannot be converted to procedural surfaces or solids. Each solid face is converted into a separate surface.

Editing NURBS surfaces

Basically, there are two ways to edit NURBS surfaces: applying the 3DEDITBAR command and editing control vertices. These can be shown, dragged to new positions, thus modeling the surface, and vertices can be added or removed. A last command allows rebuilding the surface.

The 3DEDITBAR command

The 3DEDITBAR (no alias) command allows modifying NURBS surfaces and splines by moving points and changing direction and magnitude at specific points. Besides the command line, the command is available only on the ribbon, **Surface** tab. The command starts by prompting the selection of a NURBS surface or spline:

```
Command: 3DEDITBAR
Select a NURBS surface or curve to edit: Selection
```

Then, it prompts for a point on the surface, displaying two perpendicular red lines:

```
Select point on NURBS surface. Point
```

The 3DMOVE gizmo is displayed with three grips, which is explained next. Pressing *Enter* ends the command. The **Base point** option allows selecting a different point on the surface and the **Displacement** option allows defining a new base point by projecting coordinates on the surface:

```
Select a grip on the edit bar or [Base point/Displacement/Undo/
eXit]<exit>: Selection or Enter
```

There are three grips associated to the gizmo that allow shaping the surface:

- **Square grip**: This grip moves the surface or aligns its tangent (accessed by the right-click button menu)
- **Triangle grip**: This grip allows choosing to move or to align the tangent
- **Tangent arrow grip**: This grip modifies the tangent magnitude

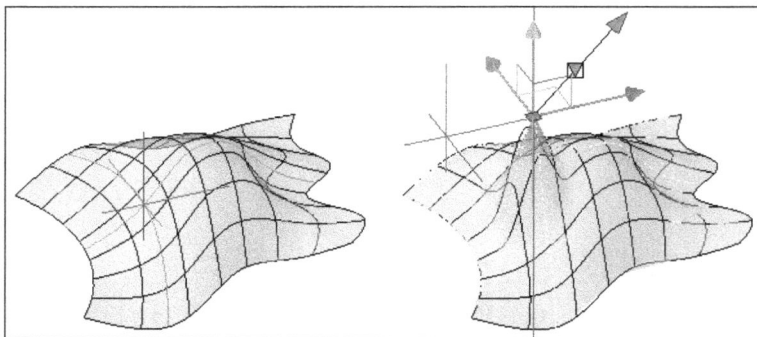

The CVSHOW command

The CVSHOW command (no alias) allows displaying the control points for selected NURBS surfaces or splines. It just prompts for the selection:

```
Command: CVSHOW
Select NURBS surfaces or curves to display control Vertices: Selection
```

The CVHIDE command

The CVHIDE command (no alias) turns off the control point's visualization for all NURBS surfaces and splines. It prompts nothing.

The CVADD command

The CVADD command (no alias) allows adding control points in the U or V direction, or directly on the surface or spline. The command prompts for the selection of a NURBS surface or spline:

```
Command: CVADD
Select a NURBS surface or curve to add control vertices: Selection
```

By default, the specified point adds a row of control vertices in U direction and the command ends. The **Direction** option allows changing to the other direction. The **insert Knots** option allows inserting knots on the surface, instead of control vertices:

```
Adding control vertices in U direction
Select point on surface or [insert Knots/Direction]: Direction
Adding control vertices in V direction
Select point on surface or [insert Knots/Direction]: Point
```

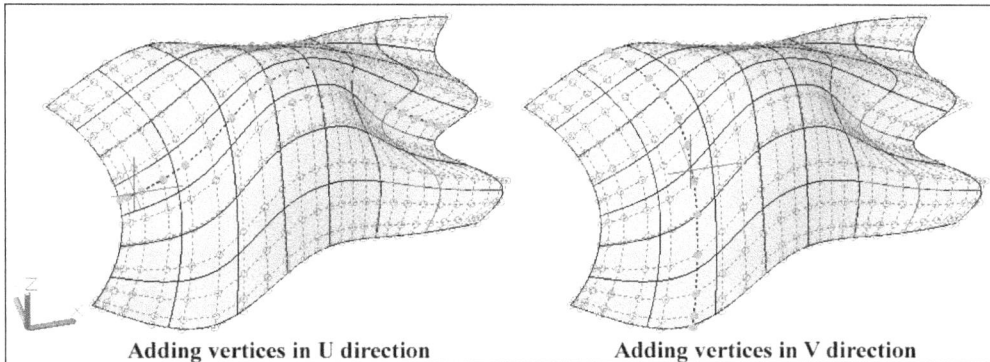

Adding vertices in U direction   **Adding vertices in V direction**

The CVREMOVE command

The CVREMOVE command (no alias) allows you to remove control vertices. The command prompts for the selection of a NURBS surface or spline:

```
Command: CVREMOVE
Select a NURBS surface or curve to remove control vertices: Selection
```

The existing rows of control vertices are displayed, in order to select the one that will be removed. The **Direction** option allows changing to the other direction:

```
Removing control vertices in U direction
Select point on surface or [Direction]: Point
```

The CVREBUILD command

The CVREBUILD command (no alias) allows rebuilding NURBS surfaces and splines. When a NURBS surface or curve has too many vertices or when we need to increase or decrease the degree, this is the command to apply. The command prompts for the selection of a surface or spline and displays a dialog box:

```
Command: CVREBUILD
Select a NURBS surface or curve to rebuild: Selection
```

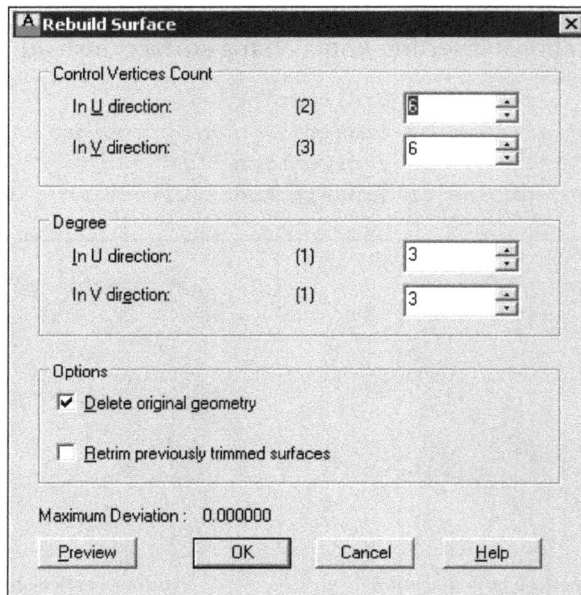

This box has the following options:

- **Control Vertices Count**: In this option, we control the number of control vertices in U and V directions.

- **Degree**: In this option, we control the surface degree in U and V directions.

- **Options**: Deleting original geometry does not maintain the original surface when sorting the command. If the original surface is trimmed (**Retrim previously trimmed surface**), this option maintains the trim on the new surface.

Exercise 12.3

We are going to create a NURBS surface to simulate a circular terrain:

1. Start a new drawing.

2. Create a circle with 40 units radius.

3. A layer, called Surface, with color at choice, should be created and activated.

4. With the PLANESURF command and the **Object** option, a planar round surface is created.

5. Applying the CONVTONURBS command, this surface is transformed into a NURBS surface.

6. To display control points, apply the CVSHOW command and select the surface. There are only four control points at the corners of the bounding rectangle.

7. The CVREBUILD command allows you to increase control points and degree. Define six control vertices in each direction.

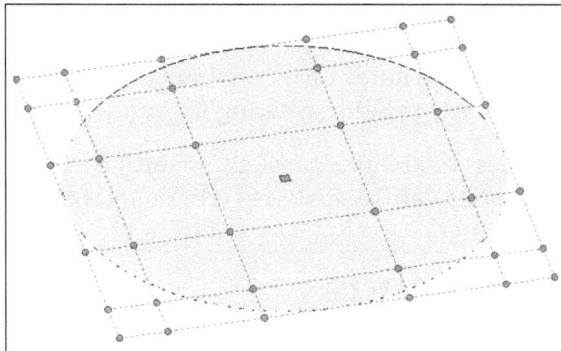

8. Now it is easy to model the surface. With **Polar** or **Ortho** turned on, move control points up or down. To pick control points, it is better to use the **2D wireframe** visual style.

9. Save the drawing with the name `A3D_12_03final.DWG`.

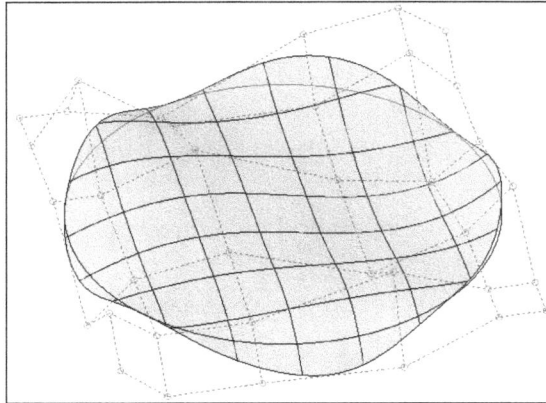

Meshes

Meshes are 3D objects that can simulate smooth or organic models. These meshes are represented by planar faces that enclose the object, called tessellation. Meshes were introduced in Version 2010 with additional commands added in Version 2011. The normal workflow for meshes is roughly creating the model and then smoothing it.

There are three possibilities for creating meshes: primitives, from linear objects, or by conversion. Several commands are available to edit or smooth meshes.

Access to commands

The mesh commands are on the ribbon, **Mesh** panel, the **Draw/Modeling** menu bar, and the **Smooth Mesh** toolbar.

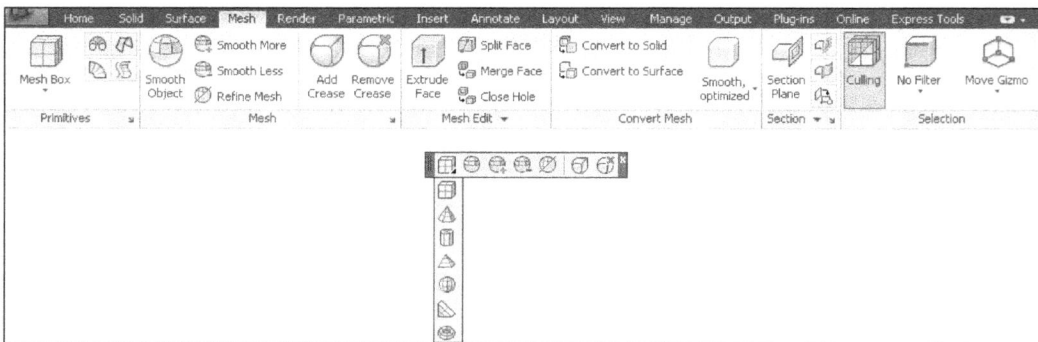

Creating primitive meshes

There is one command to create primitive meshes and another to specify default parameters for these meshes.

The MESH command

The MESH command (no alias) creates meshes with the six basic geometric shapes. The command starts by displaying the current smoothness level and prompts for the type of shape:

```
Command: MESH
Current smoothness level is set to: 0
Enter an option [Bcx/Cone/CYlinder/Pyramid/Sphere/Wedge/Torus/
SEttings] <Wedge>: Option
```

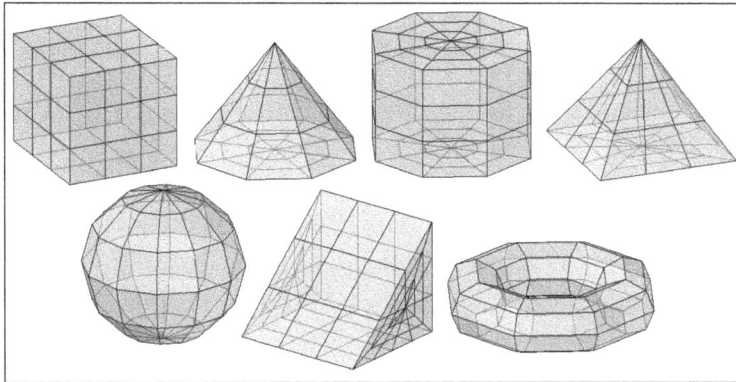

The options for this command are similar to the primitive solids:

- **Box**: This option creates a mesh with a box. By default it prompts for the first corner, the opposite corner, and height.

- **Cone**: This option creates a mesh with a cone. By default it prompts for center of base, base radius, and height.

- **CYlinder**: This option creates a mesh with a cylinder. By default it prompts for center of base, base radius, and height.

- **Pyramid**: This option creates a mesh with a pyramid. By default it prompts for center of base, base radius of circumscribed square, and height.

- **Sphere**: This option creates a mesh with a sphere. By default it prompts for center and radius.

- **Wedge**: This option creates a mesh with a wedge (box cut in diagonal). By default it prompts for the first corner, the opposite corner, and height.

- **Torus**: This option creates a mesh with a torus. By default it prompts for center, radius of torus, and radius of tube.

- **SEttings**: This option allows defining the smoothness level or tessellation (number of planar faces in each direction, also set with the next command).

The smoothness level rounds edges, thus creating smoother objects. It accepts values from 0 (no smooth) to 4 (smoothest).

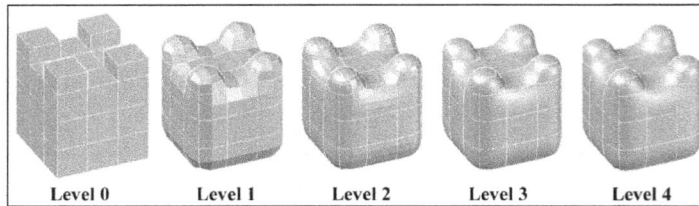

| Level 0 | Level 1 | Level 2 | Level 3 | Level 4 |

The smoothness level can be defined within the **SEttings** option, or at any time with the PROPERTIES palette.

The MESHPRIMITIVEOPTIONS command

The MESHPRIMITIVEOPTIONS command (no alias) controls the number of default tessellations when creating primitive meshes. Tessellations must be defined before objects creation; it can not be modified later.

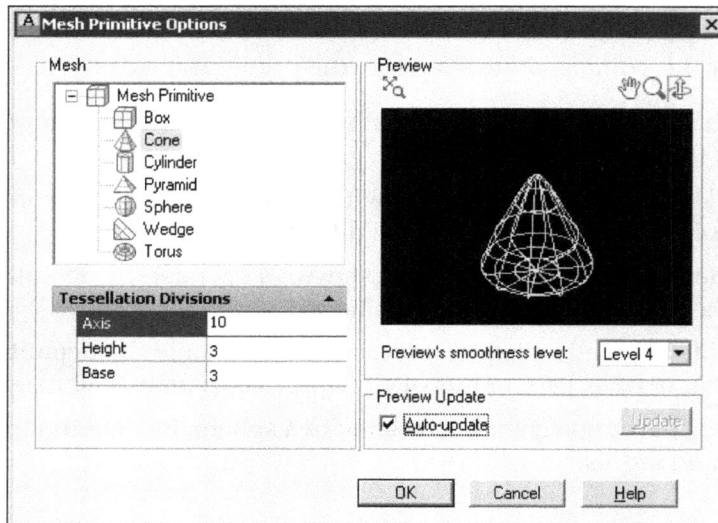

The command displays a dialog box with the following areas:

- **Mesh**: For the seven primitive meshes to be created, this option controls the default tessellation divisions in each direction
- **Preview**: This area displays the mesh primitive considering the tessellation values and also its aspect if modifying the smoothness level later
- **Preview Update**: With the help of this option, preview can be automatic or when pressing the **Update** button

Creating meshes from linear objects

Here are four commands that exist in AutoCAD for a long time. Prior to Version 2010, these commands created polyface meshes, based on 3D faces. Since that version, the MESHTYPE variable controls which type of object is created when applying the next commands. With value 1 (default), it creates meshes and with value 0, it creates polyface meshes as in older versions.

These four commands create meshes from existing linear objects. The mesh density in the first direction is controlled by the SURFTAB1 variable. The mesh density in the second direction is controlled by the SURFTAB2 variable. These values must be set before creating these meshes.

The RULESURF command

The RULESURF command (no alias) creates a mesh between two curves or one curve and one point. It is not possible to generate a mesh between a closed curve and an open one. The command just displays the current value of the SURFTAB1 variable and prompts for the selection of the first curve and the second one:

```
Command:  RULESURF
Current wire frame density:  SURFTAB1=20
Select first defining curve: Selection
Select second defining curve: Selection
```

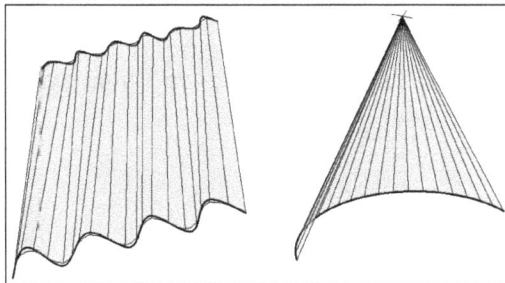

[📝 Open curves must be selected on the same side to avoid creating crossed meshes.]

The TABSURF command

The TABSURF command (no alias) creates a mesh from a curve, open or closed, and a direction vector. The command just displays the current value of the SURFTAB1 variable and prompts for the selection of the curve for the selection of the direction vector:

```
Command:  TABSURF
Current wire frame density:  SURFTAB1=20
Select object for path curve: Selection
Select object for direction vector: Selection
```

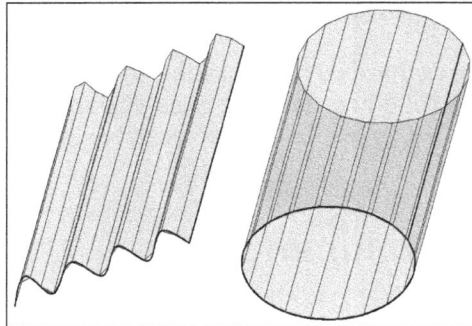

The REVSURF command

The REVSURF command (no alias) creates a revolution mesh from a curve that defines a half-crossed section, a line or polyline that defines the axis of revolution, an initial angle, and an included angle. The command starts by displaying current values of the SURFTAB1 and SURFTAB2 variables and prompts for the selection of the half-section curve:

```
Command: REVSURF
Current wire frame density:  SURFTAB1=20  SURFTAB2=6
Select object to revolve: Selection
```

The object that defines the axis of revolution is requested. We need to select an existing line or polyline, with the last axis that is defined by endpoints:

```
Select object that defines the axis of revolution: Selection
```

We define the initial angle, which is related to the section position:

```
Specify start angle <0>: Enter or value
```

We define the included angle, which by default creates a full revolution:

```
Specify included angle (+=ccw, -=cw) <360>: Enter or value
```

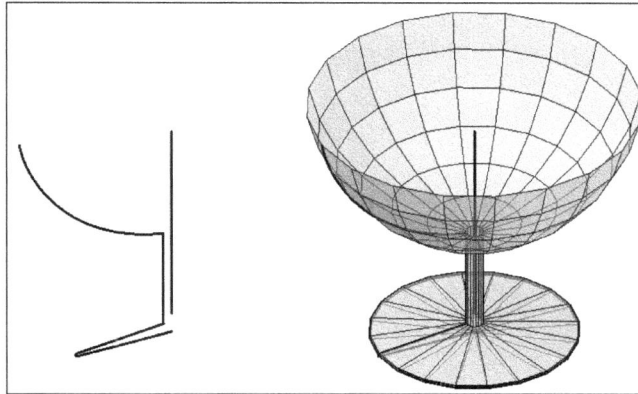

> The SURFTAB1 variable controls mesh density along revolution and the SURFTAB2 variable controls mesh density along curved parts of the original section. Straight parts have no divisions.

The EDGESURF command

The EDGESURF command (no alias) creates a mesh from four linear objects that define a closed boundary. The command starts by displaying current values of the SURFTAB1 and SURFTAB2 variables and prompts for the selection of four curves with coincident endpoints, by any order:

```
Command: EDGESURF
Current wire frame density:  SURFTAB1=20  SURFTAB2=6
Select object 1 for surface edge: Selection
Select object 2 for surface edge: Selection
Select object 3 for surface edge: Selection
Select object 4 for surface edge: Selection
```

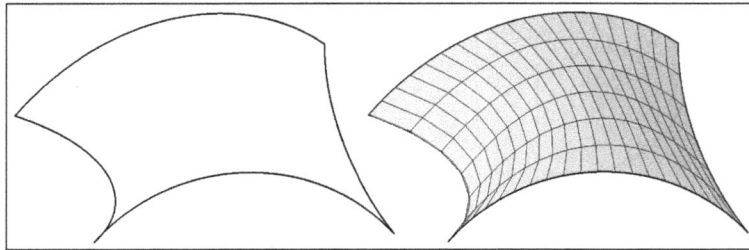

The SURFTAB1 variable is applied along the first selected linear object.

Creating meshes by converting objects

Next is a command to convert objects to meshes and another command to control conversion parameters.

The MESHSMOOTH and CONVTOMESH commands

The MESHSMOOTH command (alias SMOOTH) allows converting solids, surfaces, polyface meshes, 3D faces, regions, and closed polylines into meshes. In Version 2011, the CONVTOMESH command was introduced, but is like an alias to the MESHSMOOTH command. The conversion is made according to settings predefined in the next command.

The command may display an information box alerting for better results with primitive solids, and only prompts for object selection:

```
Command: MESHSMOOTH
Select objects to convert: Selection
```

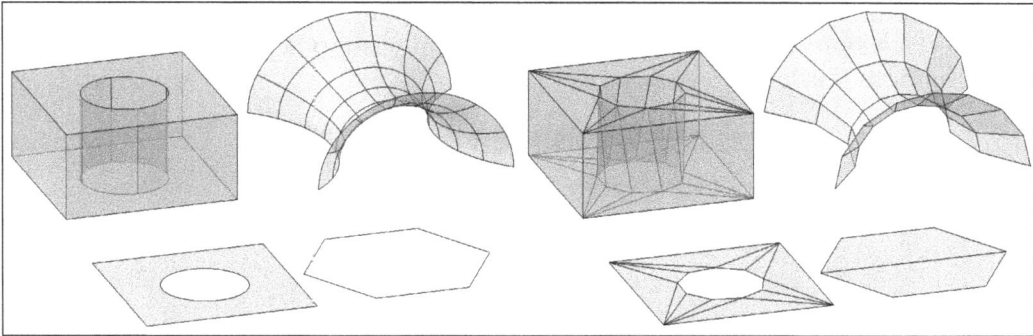

The MESHOPTIONS command

The MESHOPTIONS command (no alias) controls default settings when converting objects to meshes. The following areas are available:

- **Select objects to tessellate**: This button allows you to select objects to convert them with default settings. This option can replace the MESHSMOOTH command.

- **Mesh Type and Tolerance**: **Mesh type** can be **Smooth Mesh Optimized**, which is an optimization defined by the command, **Mostly Quads**, which is mainly creating four-sided faces, and **Triangles**, which is mainly creating three-sided faces. The following are the options available in the **Mesh Type and Tolerance** area:

    ° **Mesh distance from original faces**: This option specifies the maximum possible distance from original to mesh

    ° **Maximum angle between new faces**: This option specifies maximum angle between normal to contiguous faces

    ° **Maximum aspect ratio for new faces**: This option specifies width/height ratio to new faces

    ° **Maximum edge length for new faces**: This option specifies maximum length to face edges

- **Meshing Primitive Solids**: When converting solid primitives, this option allows you to apply predefined tessellations. The **Mesh Primitives...** button displays the **MESHPRIMITIVEOPTIONS** command box.

- **Smooth Mesh After Tessellation**: With the help of this option, after conversion, new meshes can be automatically smoothed with the chosen smoothness level.

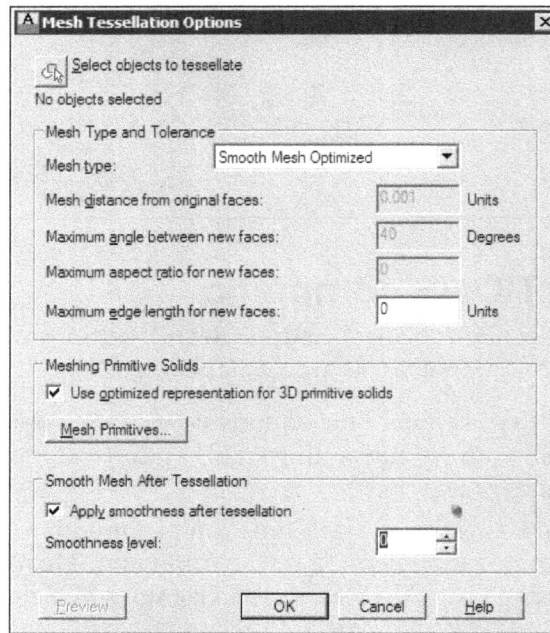

Editing meshes

Some commands are exclusive for editing meshes. However, we may select mesh faces (selected by holding down the *Ctrl* key) and apply normal editing commands, such as MOVE, ROTATE, or SCALE. Also we may apply specific 3D editing commands, such as 3DMOVE, 3DROTATE, and 3DSCALE.

When editing commands, to avoid wrong selections, it is possible to filter which mesh subobjects could be selected. The SUBOBJSELECTIONMODE variable filters the type of subobjects that can be selected, which is 0 for no filter, 1 for only vertices, 2 for only edges, and 3 for only faces.

The MESHEXTRUDE command

The MESHEXTRUDE command (no alias) allows extruding selected faces. By default, the command prompts for faces selection and extrusion height.

```
Command: MESHEXTRUDE
```

Information about the default setting concerning extrusion of consecutive faces is displayed:

```
Adjacent extruded faces set to: Join
```

The command prompts for the selection of faces. These must be selected one by one:

```
Select mesh face(s) to extrude or [Setting]: Selection
Select mesh face(s) to extrude or [Setting]: Enter
```

By default, we specify the height of extrusion:

```
Specify height of extrusion or [Direction/Path/Taper angle] <0.4000>:
Value
```

The options for this command are:

- **Setting**: This option controls if consecutive faces stay connected when smoothing or not being parallel (**Join Yes**) or being parallel (**Join No**)
- **Direction**: Similar to the EXTRUDE command, this option allows you to specify height and direction of extrusion
- **Path**: Similar to the EXTRUDE command, this option allows you to select a path and extrudes selected faces along that path
- **Taper angle**: Similar to the EXTRUDE command, this option allows you to define a taper angle for extrusion

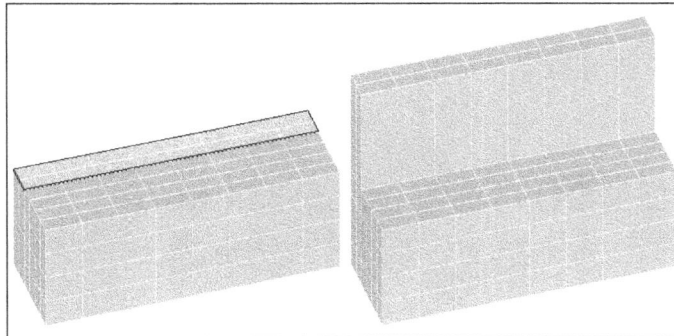

> Faces must be selected one by one; this means that it is not easy to manipulate many faces.

The MESHSCAP command

The MESHCAP command (no alias) allows you to create faces that connect edges or that close gaps. The command prompts for edges selection:

```
Command: MESHCAP
Select connecting mesh edges to create a new mesh face...
```

Instead of edges selection, the **CHain** option allows you to select one edge, and all contiguous edges are selected:

```
Select edges or [CHain]: CHain
```

The **OPtions** option specifies if the command tries to close the loop defined by all edges. **Edges** return to selection of individual edges:

```
Select edge of chain or [OPtions/Edges]: OPtions
Try to chain closed loop? [Yes/No]: <Y>: Y
```

Pressing *Enter* ends the command or the **Edges** option can be used to select more edges:

```
Select edge of chain or [OPtions/Edges]: Enter
```

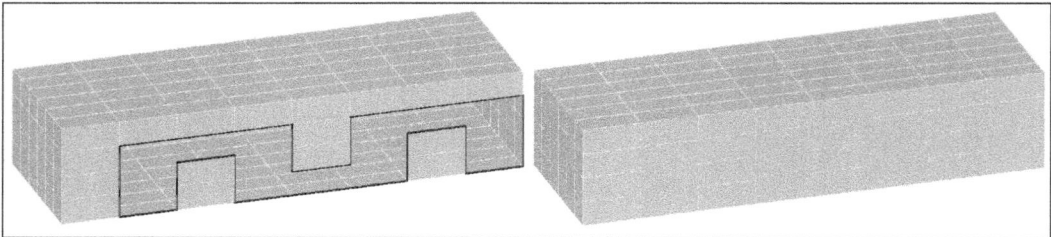

The MESHSMOOTHMORE command

The MESHSMOOTHMORE (alias MORE) command increases the smoothness of the selected mesh one level. It only prompts for meshes selection.

> A more efficient way to modify the smoothness level is by applying the PROPERTIES palette.
>
> If this command is applied to objects that are not meshes, but can be converted, it prompts for conversion confirmation.

The **MESHSMOOTHLESS** command

The MESHSMOOTHLESS command (alias LESS) decreases the smoothness of the selected mesh one level. It only prompts for selection of meshes.

The **MESHCREASE** command

The MESHCREASE command (alias CREASE) allows you to define sharpen or creased edges that are not affected by increasing the smoothness level. The command prompts for mesh subobjects selection:

```
Command: MESHCREASE
Select mesh subobjects to crease: Selection
Select mesh subobjects to crease: Enter
```

We can define a maximum smoothness level from which crease will not be maintained. By default, **Always** maintains crease despite the smoothness level:

```
Specify crease value [Always] <Always>: Enter
```

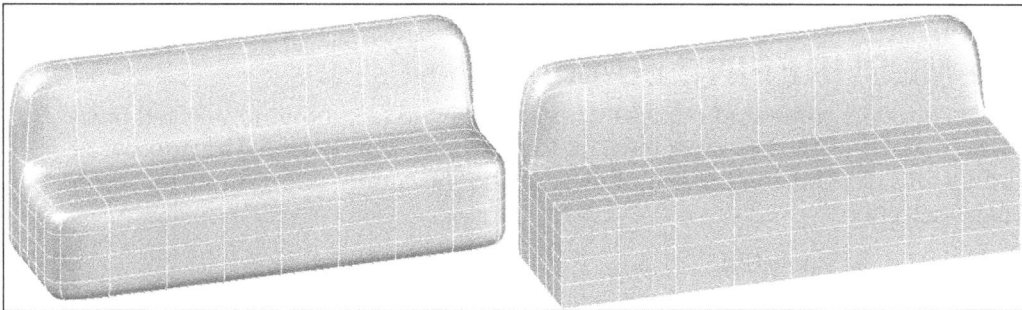

> The PROPERTIES palette allows modifying crease to selected subobjects.

The MESHUNCREASE command

The MESHUNCREASE command (alias UNCREASE) allows you to remove crease edges, defined by the previous command. The command just prompts for creases to be removed:

```
Command: MESHUNCREASE
Select crease to remove: Selection
Select crease to remove: Enter
```

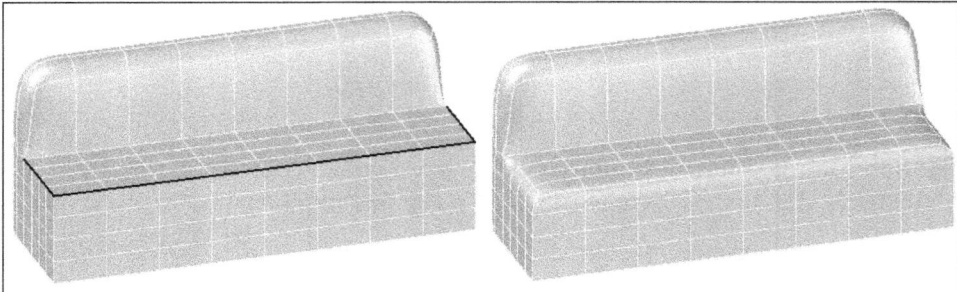

The MESHREFINE command

The MESHREFINE command (alias REFINE) allows you to subdivide selected faces or whole meshes. Meshes must have, at least, smoothness level 1. The number of subdivisions depends on the smoothness level. The command prompts for meshes or faces selection:

```
Command: MESHREFINE
Select mesh object or face subobjects to refine: Selection
Select mesh object or face subobjects to refine: Enter
```

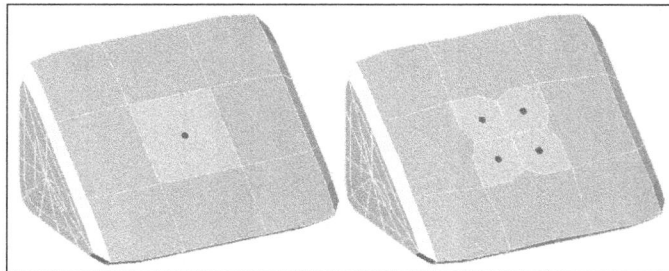

The MESHMERGE command

The MESHMERGE command (no alias) joins contiguous faces into one. The command just prompts for face selection:

```
Command: MESHMERGE
Select adjacent mesh faces to merge: Selection
Select adjacent mesh faces to merge: Enter
```

The MESHCOLLAPSE command

The MESHCOLLAPSE command (no alias) merges vertices from selected faces or edges. The command just prompts for the selection of one mesh face or mesh edge:

```
Command:  MESHCOLLAPSE
Select mesh face or edge to collapse: Selection
```

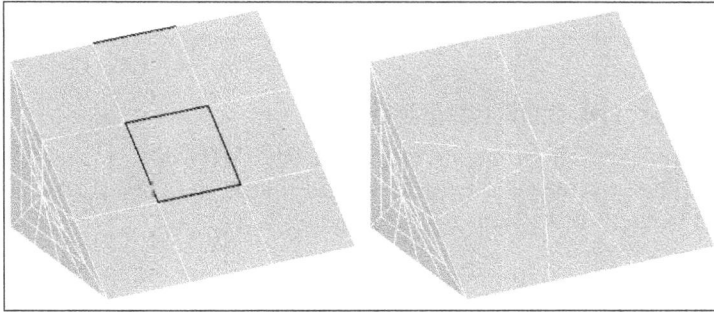

The MESHSPLIT command

The MESHSPLIT command (alias SPLIT) allows dividing a mesh face in two faces. The command prompts for a face selection and two cutting vertices on edges:

```
Command: MESHSPLIT
Select a mesh face to split: Selection
```

When passing over an edge, a small knife symbol is added to the cursor. The **Vertex** option limits point selection to vertices:

```
Specify first split point on face edge or [Vertex]: Point
Specify second split point on face edge or [Vertex]: Point
```

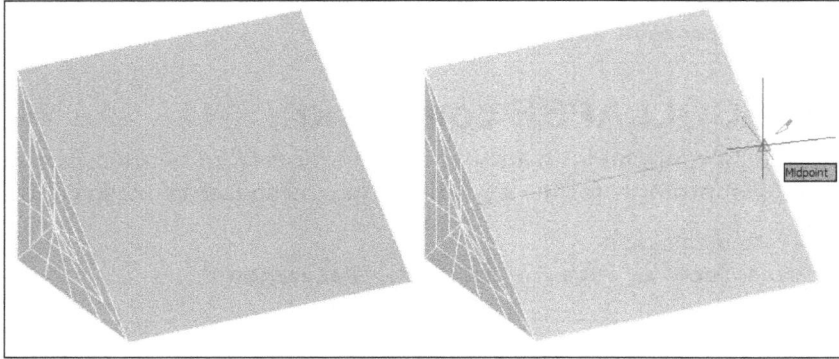

The MESHSPIN command

The MESHSPIN command (no alias) changes the edge between two contiguous triangular faces. The command prompts for the selection of the first and the second triangular mesh faces that is contiguous to the first one:

```
Command: MESHSPIN
Select first triangular mesh face to spin: Selection
Select second adjacent triangular mesh face to spin: Selection
```

Polyface meshes

Polyface meshes were the first 3D objects in AutoCAD. These meshes are based on 3D faces defined by three or four vertices. Only the 3DFACE command is important, as introduced and used in *Chapter 5, 3D Primitives and Conversions*. Polyface meshes still have one advantage related to other surfaces and meshes: vertices can be stretched easily.

Creating polyface meshes

Besides the important 3DFACE command, some others are available, but mainly used for programming. None of these commands have icons.

If the MESHTYPE variable has value 0; the RULESURF, TABSURF, REVSURF, and EDGESURF commands, which were presented before, create polyface meshes. The result of exploding polyface meshes are 3D faces.

These commands have no **Undo** option. An error in specifying a vertex requires command cancelation or posterior modification.

The 3DFACE command

The 3DFACE command (alias 3F) creates faces defined by three or four vertices. A single command application allows creating several contiguous faces:

```
Command: 3DFACE
```

The command prompts for a sequence of points that are defined in clockwise or counterclockwise directions. The **Invisible** option allows turning invisible points to the next edge:

```
Specify first point or [Invisible]: Point1
Specify second point or [Invisible]: Point2
Specify third point or [Invisible] <exit>: Point3
Specify fourth point or [Invisible] <create three-sided face>: Point4
Specify third point or [Invisible] <exit>: Point5
Specify fourth point or [Invisible] <create three-sided face>: Point6
Specify third point or [Invisible] <exit>: Point7
Specify fourth point or [Invisible] <create three-sided face>: Enter
Specify third point or [Invisible] <exit>: Enter
```

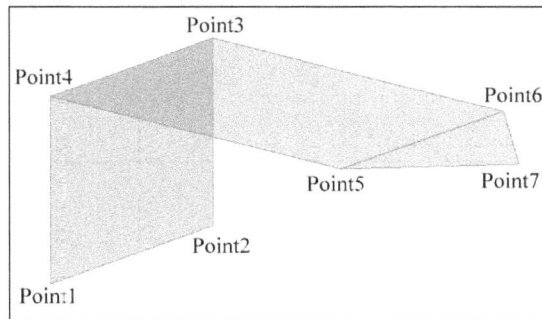

The 3DMESH command

The 3DMESH command (no alias) creates a polyface mesh by specifying how many vertices exist in both directions and each vertex coordinates. This command is mainly used for programming meshes from a list of point coordinates:

```
Command: 3DMESH
```

We define the number of vertices in the first direction:

```
Enter size of mesh in M direction: 2
```

Then we define the number of vertices in the second direction:

```
Enter size of mesh in N direction: 4
```

Finally, we define each vertex location:

```
Specify location for vertex (0, 0): Point
Specify location for vertex (0, 1): Point
Specify location for vertex (0, 2): Point
Specify location for vertex (0, 3): Point
Specify location for vertex (1, 0): Point
Specify location for vertex (1, 1): Point
Specify location for vertex (1, 2): Point
Specify location for vertex (1, 3): Point
```

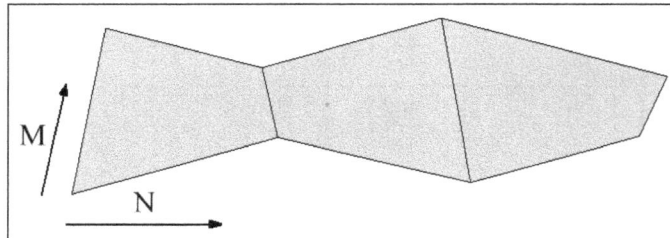

The PFACE command

The PFACE command (no alias) creates polyface meshes composed by faces that can have more than four edges. The command starts by prompting for the location of all vertices and then prompts for the vertices that belong to face 1, face 2, and so on. When specifying a new face, this can have a different color or layer.

The 3D command

The 3D command allows you to create polyface meshes corresponding to basic volume shapes, such as boxes, wedges, pyramids, meshes, torus, cones, spheres, and hemispheres.

> This command is defined by an AutoLISP program, called 3D.LSP. Since Version 2012, this program is not included, but it is easy to find it in previous versions or over the Internet and load it into AutoCAD.

Editing polyface meshes

The STRETCH command is a great tool to modify polyface meshes. It's as easy as selecting vertices with crossing mode and moving them to a new position.

The PEDIT command, when selecting a polyface mesh obtained with 3DMESH, 3D, RULESURF, TABSURF, REVSURF, and EDGESURF commands, has specific options:

- **Edit vertex**: This allows editing vertices, basically by selecting one and moving it (it is easier with the STRETCH command)
- **Smooth surface**: This smoothes the polyface mesh
- **Desmooth**: This cancels the effect of the **Smooth** option
- **Mclose/Mopen**: This closes or opens the mesh in the M direction
- **Nclose/Nopen**: This closes or opens the mesh in the N direction

Surface analysis

There are three commands, since Version 2011, that perform surface analysis and another activity to adjust options. These analysis are useful when needing to check continuity, curvature, and draft angles, like in the mould industry.

These commands can not be used with the **2D wireframe** visual style and are only available on the ribbon, **Surface** panel.

Types of analysis and commands

AutoCAD considers three types of analysis, which can be applied to surfaces or solids:

- **Zebra analysis**: Surfaces are analyzed about their continuity by projecting parallel black and white stripes onto the model (default), where surfaces meet and stripes allow analyzing curvature and tangency
- **Curvature analysis**: Surfaces are analyzed about their curvature with a color gradient
- **Draft analysis**: Surfaces are analyzed about how easy they would be separated from the mould

The ANALYSISZEBRA command

The ANALYSISZEBRA command (alias ZEBRA) prompts for the selection of solids and surfaces and displays stripes on surfaces. These stripes allow you to detect nontangencies and curvature changes:

```
Command: ANALYSISZEBRA
Select solids, surfaces to analyze or [Turn off]: Selection
```

The ANALYSISCURVATURE command

The ANALYSISCURVATURE command (no alias) prompts for the selection of solids and surfaces and displays a color gradient. If gaussian curvature is positive (convex), the color is green, and if it is negative (concave), the color is blue:

```
Command: ANALYSISCURVATURE
Select solids, surfaces to analyze or [Turn off]: Selection
```

The ANALYSISDRAFT command

The ANALYSISDRAFT command (no alias) prompts for the selection of solids and surfaces and displays a color gradient. This color gradient evaluates how easy it will be to separate the part from the mould:

```
Command: ANALYSISDRAFT
Select solids, surfaces to analyze or [Turn off]: Selection
```

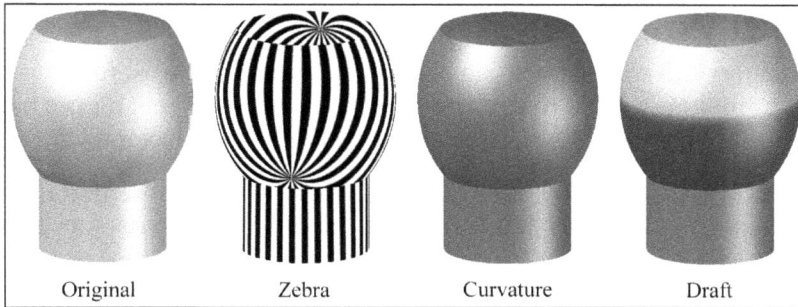

| Original | Zebra | Curvature | Draft |

The ANALYSISOPTIONS command

The ANALYSISOPTIONS command (no alias) controls options for all analysis types. The command displays a dialog box with three tabs. In all, it is possible to select the objects to analyze and clear the analysis:

- **Zebra**: This tab controls all options related to zebra analysis, namely stripes angle, the type between cylindrical or spherical, stripes size, and colors
- **Curvature**: This tab controls all options related to curvature analysis, namely display style and color range
- **Draft Angle**: This tab controls color range for draft angle analysis

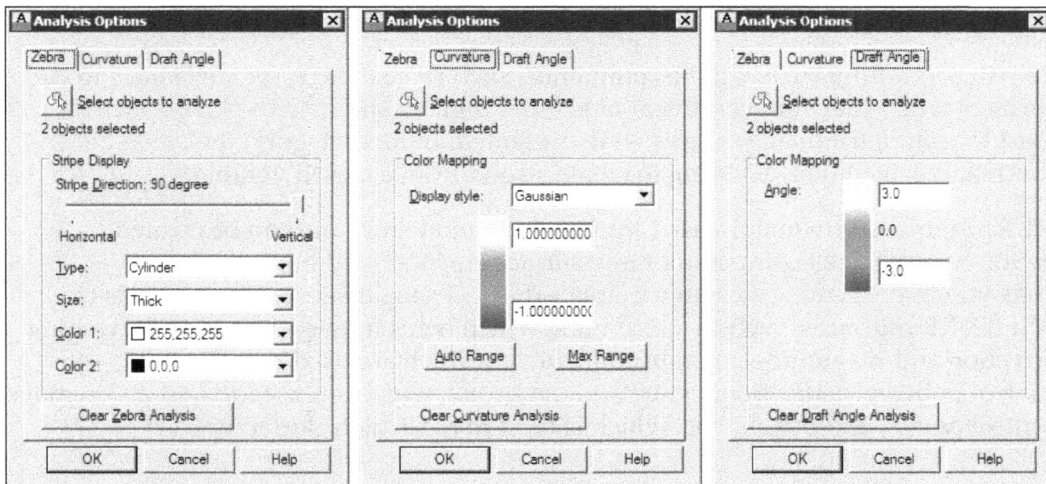

Summary

Besides modeling with solids, for some projects, surfaces can be useful. This chapter includes all commands related to surfaces and meshes, objects that do not have thickness. We started by identifying all categories that were developed in the chapter.

Besides the EXTRUDE, REVOLVE, SWEEP, and LOFT commands (presented in *Chapter 4, Creating Solids and Surfaces from 2D*) and the EXTRUDE and CONVTOSURFACE commands (presented in *Chapter 5, 3D Primitives and Conversions*), several other procedural surfaces commands are available, which are as follow:

- SURFNETWORK: This command is used to create a surface from two sets of linear objects
- SURFBLEND: This command is used to create a surface connecting two existing surfaces
- SURFPATCH: This command is used to create a surface that closes a hole or an opening
- SURFEXTEND: This command is used to create a surface by extending existing surfaces from selected edges
- SURFOFFSET: This command is used to create surfaces parallel to the selected surfaces
- SURFFILLET: This command is used to create a fillet surface between two existing surfaces

To edit procedural surfaces, the commands, such as SURFTRIM, which is used to cut surfaces with other surfaces, linear objects, or regions and SURFUNTRIM, which is used to restore trimmed surfaces were presented. In this category, the SURFSCULP command is included, allowing to create a solid from a closed volume.

NURBS surfaces are mainly used for free-form modeling, and can be created by the CONVTONURBS command or any surface creation method, having the SURFACEMODELINGMODE system variable value 1. There are several commands to edit NURBS surfaces, such as 3DEDITBAR, which helps moving points and changing direction and magnitude, CVSHOW and CVHIDE, which are used to turn on and off control points visualization, CVADD and CVREMOVE, which are used to add and remove control points, and CVREBUILD, which is used to recalculate surfaces.

Meshes are 3D objects that can simulate smooth or organic models, which can be created in three ways: the MESH command, which has options defined by the MESHPRIMITIVEOPTIONS command, RULESURF, TABSURF, REVSURF, and EDGESURF commands, which are used to create existing objects from linear objects, MESHSMOOTH (or CONVTOMESH) command, which is used to convert existing objects and conversion options set with MESHOPTIONS command. There are several commands to edit meshes, which are as follows:

- MESHEXTRUDE: This command is used to extrude selected faces
- MESHCAP: This command is used to create faces that connect edges or that close gaps
- MESHSMOOTHMORE and MESHSMOOTHLESS: These command are used to increase and decrease smoothness level
- MESHCREASE and MESHUNCREASE: These command are used to define and remove creased edges
- MESHREFINE: This command is used to subdivide selected faces or whole meshes
- MESHMERGE: This command is used to join contiguous faces into one
- MESHCOLLAPSE: This command is used to merge vertices
- MESHSPLIT: This command is used to divide mesh faces
- MESHSPIN: This command is used to rotate the edge between two contiguous triangular faces

Polyface meshes are based on 3D faces defined by three or four vertices, and were the first 3D objects in AutoCAD. To create polyface meshes there are commands, such as 3DFACE, which is used to create faces defined by three or four vertices, 3DMESH, which is used mainly for programming and to create a polyface mesh by specifying how many vertices exist in both directions and each vertex coordinates, PFACE, which is used to create polyface meshes composed by faces that can have more than four edges, and 3D, which is used to create polyface meshes corresponding to basic volume shapes. The STRETCH command allows moving vertices of polyface meshes.

Finally, four commands were presented, related to surface analysis, namely: ANALYSISZEBRA, which is used to analyze surfaces about continuity, ANALYSISCURVATURE, which is used to analyze surfaces about its curvature, ANALYSISDRAFT, which is used to analyze surfaces about how easy would they be separated from mould, and ANALYSISOPTIONS, which is used to control options for all analysis types.

Final Considerations

In this appendix, we include creating simple animations representing walkthroughs or see-around import and export file formats, advices for exporting from AutoCAD to 3ds Max and Revit, and development clues for 3D modelers.

The topics covered in this appendix are:

- How to create simple animations
- Preparing a model to export to other applications
- How to develop in 3D

Advanced concepts and clues for development

Until now, we have seen everything for creating great models and obtaining excellent rendered images. In this appendix we see how to create simple animations, how to prepare a model and export it, and also suggest how to progress in 3D.

Animation

AutoCAD has limited capabilities for creating animations, which are described here. With the next command, we may associate a camera and its target to points or paths and animate it.

The ANIPATH command

The ANIPATH command (no alias) allows simulating a walkthrough or a see-around. This command displays the **Motion Path Animation** dialog box where a camera can be associated to a point or a path, and the camera target can also be associated to a point or a path. Paths are linear objects, which were previously created.

This dialog box has the following areas and options, as displayed in the next screenshot:

- **Camera**: This command automatically creates a camera that can be associated to a point (fixed camera) or a path (moving camera). The button hides the box temporarily, allowing you to mark a point or to select a linear object. After selection, a box to introduce a name is displayed.

- **Target**: This camera target can also be a point or it follows a path, which is eventually the same as that of the camera's. The button allows you to select the path or a point.

- **Animation Settings**: **Frame rate (FPS)** defines how many frames per second is the animation velocity. **Number of frames** controls how many frames will have the animation. **Duration (seconds)** controls the animation duration. Modifying duration and the number of frames adjust accordingly. The **Visual style** list controls which visual style or render preset will be used in the animation. **Format** allows you to choose the animation type between AVI, MOV, MPG, or WMV. **Resolution** allows you to choose the resolution, the maximum being **1024 x 768**. With **Corner deceleration** checked, the animation slows down on the corners. **Reverse** reverses the animation.

- **When previewing show camera preview**: With this option unchecked, the camera or target animates on the viewport, but no preview window having the camera view is displayed when pressing the **Preview** button.

- **Preview...**: This button displays an animation preview.

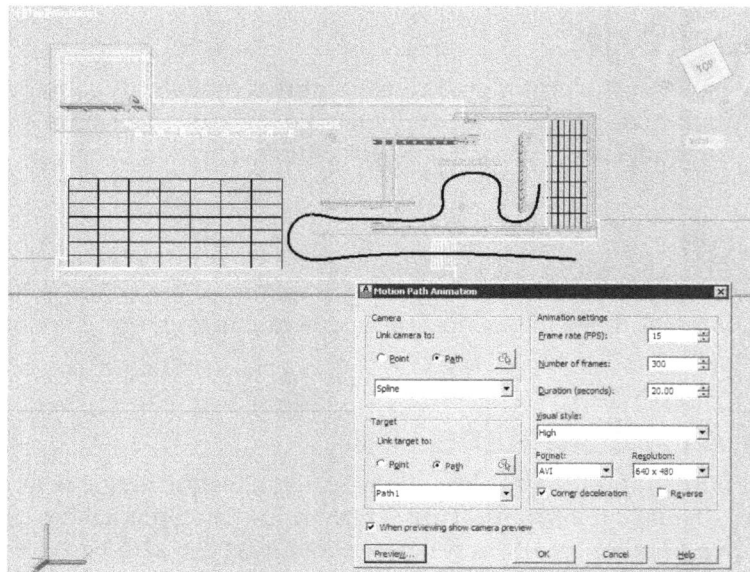

- **OK**: This button creates the animation. A **File** dialog box is displayed and then the animation process starts. Depending on model complexity, materials, lighting, animation resolution, and number of frames, the calculation may take a huge amount of time. Basically, the time taken is similar to the time of a single render with the same resolution times the number of frames.

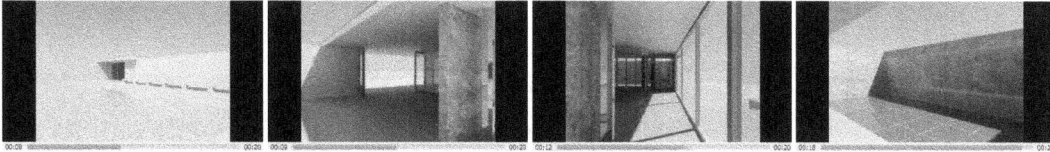

> The ANIPATH command has several severe limitations: it is not possible to associate an existing camera to create the animation, it is not possible to control the camera's parameters, such as lens length or field of view, and it is not possible to control output quality, such as selecting an animation codec.

Connecting to other programs

Often we need to import or export our model to other programs, such as 3ds Max or Revit. As we are dealing with different types of files, some care must be taken.

Typical 3D import and export formats

It is quite simple to connect to AutoCAD. Besides the AutoCAD DWG file type that most CAD systems can export or import, the program accepts several 3D file types:

- **SAT**: These are the ACIS solid modeling files
- **FBX**: These are the 3D standard files that allow sharing 3D models between Autodesk software, such as Revit
- **DGN**: These are the Microstation files
- **IGES and IGS**: These are the standard neutral files
- **STL**: These files are only for exporting stereo lithography files, which are used for prototyping machines

Additionally, AutoCAD can import the following file types:

- **3DS**: These are the old files of 3D Studio (MS-DOS versions)
- **IPT and IAM**: These are the Autodesk Inventor parts and also the assembly files
- **MODEL, SESSION, EXP, DLV3, CATPART, and CATPRODUCT**: These are the CATIA files
- **IJ**: These are the JT solid modeling files
- **X_B, X_T**: These are both Parasolid binary and Parasolid text files
- **PRT, ASM, G, and NEW**: These are the Pro/Engineer files
- **3DM**: These are the Rhino files
- **PRT, SLDPRT, ASM, and SLDASM**: These are the SolidWorks files
- **STE, STP, and STEP**: These are the STEP files

The IMPORT command (alias IMP) allows importing several file types. The EXPORT command (alias EXP) allows exporting the model or selected objects to several file types.

Exporting to 3ds Max

3ds Max and 3ds Max Design are the 3D Autodesk software for creating high-quality images and animations. Many users prefer to model in AutoCAD and then export to 3ds Max for applying materials, lights, defining animations, and render. Many others want to import a 2D drawing and create the 3D model on top of it.

The best file formats to communicate with 3ds Max are DWG and FBX.

There are some basic precautions that must be taken when working on an AutoCAD model if this is going to be imported to 3ds Max:

- The model must not have lost objects far away. The ZOOM extents before saving is always a good advice.
- It is also wise to purge the drawing for not taking useless information.
- The model must be placed near the origin of the world coordinate system. When opening in 3ds Max, this program places the model that is related to the world coordinate system, which is independent of the current UCS when saving the model in AutoCAD. If objects are too far from the origin, we will have huge problems in Max.
- It is completely forbidden to have overlapping objects.

- The AutoCAD model should have a good layer distribution, as the normal method in Max is to import each layer's objects as one Max object (as editable spline or editable mesh). As applying materials in Max is normally done by an object. Also if layer distribution contemplates this, it will be much easier to apply materials in Max.

Exporting to Revit

Revit is a 3D Autodesk software suite encompassing the following three programs:

- **Architecture**: This is used to project 3D architecture models
- **MEP**: This is used to project mechanical, electrical, and plumbing
- **Structure**: This is used to project building structural elements

It is possible to import 3D models in Revit and convert them into Revit objects, such as mass elements.

The best file formats to communicate with Revit are DWG and SAT.

We present some tips to prepare AutoCAD models:

- One layer must be given to each material (in Revit, materials are applied to layers, identified by object styles).
- Units must be consistent.
- An AutoCAD object's properties, such as color and linetype, must be ByLayer, so these can be modified in Revit. If objects have explicit colors or linetypes, it is not possible to modify them.
- AUDIT and PURGE should be performed before saving or exporting to SAT.

Clues for development

To be an excellent 3D modeler in AutoCAD, it is important to know all commands and concepts well, but the most important is practice, practice, and practice.

Here are some final advices for development:

- Before starting in 3D, the model must be understood well.
- All model dimensions must be known or easily decided (if we are creating something new). We may consider creating some volumes just to decide general dimensions.
- Where to start a 3D model must be planned well. Frequently a wrong start represents hours or days of lost work.

- Output is also important as it may influence the modeling. For instance, if it is for rendering, it may not need important details in far parts of the model; when creating models for prototyping in stereo lithography, the model must be a single mesh and cheaper to produce if hollowed.

- If the project includes 3D blocks, such as furniture or equipments, an Internet search may be performed, or the manufacturer may be contacted. These blocks, if found with a different file format, may be converted or imported.

- All materials must be selected to know at least that all parts have different materials. Distinct layers must be applied to all parts with different materials.

- Projects must be saved frequently in specific and organized folders. Some backups or files that are currently in use must be maintained so that the current model file can be recovered if it gets corrupted.

- Complex models should not be avoided. It's with challenging models that we progress and learn how to extract the maximum from AutoCAD.

- External references (xrefs) may be considered in order to split complex models.

- Free time can be used to create complex models; we may launch challenges to ourselves.

- The AutoCAD workspace can be customized. Instead of relying on the ribbon, a more efficient and fast way to work is configuring shortcuts or creating specific toolbars.

- The Internet, namely forums and blogs, are great sources of information and must be consulted regularly. We learn a lot with others' questions and comments; we may also ask for advice or support.

Index

P

[PACKT]
PUBLISHING

Thank you for buying
Autodesk AutoCAD 2013
Practical 3D Drafting and Design

About Packt Publishing

Packt, pronounced 'packed', published its first book "*Mastering phpMyAdmin for Effective MySQL Management*" in April 2004 and subsequently continued to specialize in publishing highly focused books on specific technologies and solutions.

Our books and publications share the experiences of your fellow IT professionals in adapting and customizing today's systems, applications, and frameworks. Our solution based books give you the knowledge and power to customize the software and technologies you're using to get the job done. Packt books are more specific and less general than the IT books you have seen in the past. Our unique business model allows us to bring you more focused information, giving you more of what you need to know, and less of what you don't.

Packt is a modern, yet unique publishing company, which focuses on producing quality, cutting-edge books for communities of developers, administrators, and newbies alike. For more information, please visit our website: www.packtpub.com.

Writing for Packt

We welcome all inquiries from people who are interested in authoring. Book proposals should be sent to author@packtpub.com. If your book idea is still at an early stage and you would like to discuss it first before writing a formal book proposal, contact us; one of our commissioning editors will get in touch with you.

We're not just looking for published authors; if you have strong technical skills but no writing experience, our experienced editors can help you develop a writing career, or simply get some additional reward for your expertise.

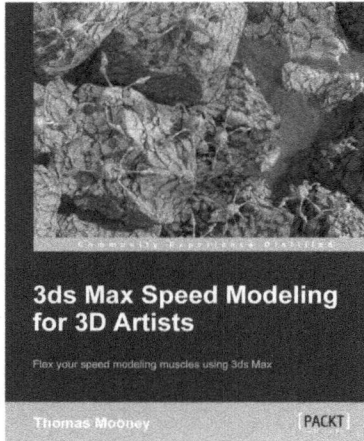

3ds Max Speed Modeling for 3D Artists

ISBN: 978-1-84969-236-6   Paperback: 422 pages

Flex your speed modeling muscles using 3ds Max

1. Learn to speed model in 3ds Max, with an emphasis on hard surfaces

2. Up to date coverage, covering 3ds Max 2013 features

3. Focused explanations with step-driven practical lessons balance learning and action

FreeCAD [Instant]

ISBN: 978-1-84951-886-4   Paperback: 70 pages

Solid Modeling with the power of Python

1. Learn something new in an Instant! A short, fast, focused guide delivering immediate results.

2. Packed with simple and interesting examples of python coding for the CAD world.

3. Understand FreeCAD's approach to modeling and see how Python puts unprecedented power in the hands of users.

4. Dive into FreeCAD and its underlying scripting language.

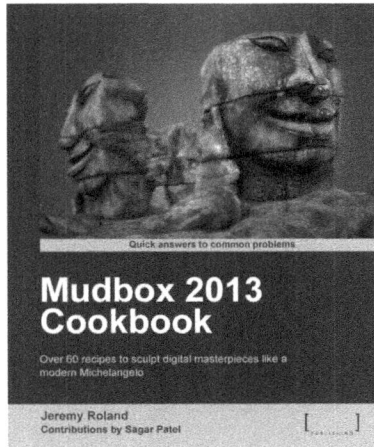

Mudbox 2013 Cookbook

ISBN: 978-1-84969-156-7   Paperback: 260 pages

Over 60 recipes to sculpt digital masterpieces like a
modern Michelangelo

1. Create amazing, high detail sculpts for games,
   movies, and more

2. Extract high resolution texture maps to use on
   your low poly 3d models

3. Create terrain that you can walk on in a
   virtual world

4. Learn professional tricks that will improve your
   workflow

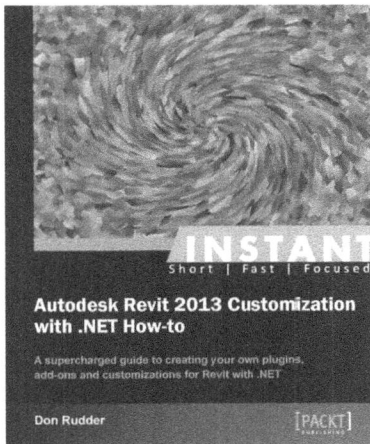

Instant Autodesk Revit 2013 Customization with .NET How-to [Instant]

ISBN: 978-1-84968-842-0   Paperback: 82 pages

A supercharged guide to creating your own plugins,
add-ons and customizations for Revit with .NET

1. Learn something new in an Instant! A short,
   fast, focused guide delivering immediate
   results.

2. Master the fundamentals of programing with
   the Autodesk Revit 2013 API

3. Customize your own ribbon controls according
   to personal preference

4. Save time and effort by learning how to
   manipulate elements and their data

Please check **www.PacktPub.com** for information on our titles